Innovations in Astronomy

The sciences move fast—
and ABC-CLIO's Innovations in Science series can help you keep pace. Each title provides an overview of the events, scientists, and innovations that have shaped the development of a particular field of science during the past hundred years. Well suited to student research, or just for satisfying the curious, these accessible handbooks for the nonspecialist highlight the scientific breakthroughs of the twentieth century and the prospects for the twenty-first.

Titles in This Series
Innovations in Astronomy
Innovations in Biology
Innovations in Earth Sciences

INNOVATIONS IN SCIENCE

Innovations in Astronomy

Overview by A. W. Jones

ABC-CLIO

Santa Barbara, California
Denver, Colorado
Oxford, England

Library of Congress Cataloging-in-Publication Data

Innovations in astronomy.
    p. cm. — (Innovations in science)
   Includes bibliographical references.
   1. Astronomy.   I. Series
QB43.2.I56   1999   520'.9'04—dc21   99-27927

ISBN 1-57607-114-6 (alk. paper)

05 04 03 02 01 00 99 10 9 8 7 6 5 4 3 2 1 (cloth)

ABC-CLIO, Inc.
130 Cremona Drive, P.O. Box 1911
Santa Barbara, California 93116-1911

This book is printed on acid-free paper ∞.

Contents

Preface

There is never a right time to take stock of astronomy, so the turn of the century is as good a moment as any. The problem, if it is a problem, is that the unparalleled scientific advances and the unremitting pace of discovery of the twentieth century show no sign of letting up. Science does not pause for us to take stock—we have to do it on the fly.

So this book does not set out to be a definitive account of the state of astronomy. Rather, it will give the reader the necessary grounding to pursue his or her own studies in astronomy into the twenty-first century. The chronology of astronomy in chapter 2 offers a year-by-year account of important advances since 1903. Here you will learn how astronomy has unfolded since the turn of the twentieth century and be able to trace the many strands of discovery as they unwind through the years.

In science it has been customary to associate advances in knowledge with individual scientists—one thinks of Darwin, Einstein, Watson, and Crick—so this book includes in chapter 3 a collection of short biographies of astronomers and scientists in closely related fields who have contributed in some important way to their disciplines in the twentieth century. Yet in astronomy, as in other disciplines, the luminaries seem to shine less intensely toward the end of the century than at the beginning. Astronomy is now "big science," almost wholly reliant on state funding, and the days when a determined individual could dominate a field are almost gone—there are no more Edwin Hubbles.

Parallel trends are at work in the institutions of astronomy. In 1900 a mere handful of institutions were doing serious research in astronomy, and they were essentially all in Europe and North America. Today, such is the proliferation and collaborative nature of astronomical research that all parts of the globe are involved.

This flowering of research groups has been made possible by a subtle but powerful change in the practice of astronomy; namely the separation of the observing sites—the physical observatory and the instruments it contains—from the institution conducting research at that observatory. No longer is astronomical investigation limited to those organizations who possess their own observatories. The provision of communal telescopes—by consortia of universities, government agencies, and international organizations—has allowed many small research groups to spring up all over the world. The European Southern Observatory, run by a consortium of eight countries, is an outstanding example of communal provision. The Hubble Space Telescope is another. (Chapter 4 lists details of some of the most important observatories and telescopes.) Many of these observatories and research and international organizations have their own Web sites and these are listed, together with a brief description, in chapter 6.

One obstacle to the public appreciation of any science is the specialized language. So, chapter 7, a dictionary containing more than six hundred astronomical terms and concepts, from aberration of starlight to zodiacal light, should help you discover the meaning of unfamiliar terms and should inform and enrich your reading. If, through reading this book, you would like to discover more, chapter 5 gives you the details of publications covering different and fascinating aspects of astronomy. Finally, this book includes as an appendix several quick-reference statistical charts followed by a subject index.

But where should you start your study of astronomy? To orient the reader, the book begins with an overview (chapter 1) of the field since 1900, starting with the tools of the astronomer and ending with recent discoveries about the origin of the universe. Here are the big ideas of the twentieth century, and the seed corn for the twenty-first.

Innovations in Astronomy

Part I

1

Overview

At the beginning of the twentieth century, the universe was a rather small place. The distances to some of the nearer stars were known, but astronomers had little idea of the overall size of the galaxy and no inkling that there might be other galaxies beyond our own. We now have a far grander conception of our place in the universe. Our sun is one of hundreds of billions of stars in the galaxy, and our galaxy is but one of hundreds of billions of galaxies grouped in clusters throughout the universe.

Such has been the explosive growth of astronomical knowledge in the twentieth century that no short overview can hope to do more than high-light some of the more significant milestones along the way or acknowl-edge more than a handful of the many scientists who have contributed to that knowledge. But we now know how stars work, how they are born, how they live, and how they die. Many exotic kinds of objects—quasars, pulsars, black holes—have been discovered that could not have been imag-ined a century ago. In our own solar system, the planets are no longer shimmering disks seen through telescopes, but whole worlds like Earth, with their own geology, climate, and prehistory. Most remarkable, we know that the universe began with a Big Bang between nine billion and fifteen billion years ago and has been expanding ever since.

The Tools of the Astronomer

Our dramatic change in perception of the universe has come about through several influences, but mention must first be made of the great strides in technology that have characterized innovation in astronomy.

At the beginning of the twentieth century, the astronomer's window on the universe was the narrow one of visible light using optical telescopes. At the end, the whole of the electromagnetic spectrum—in order of in-creasing wavelength: gamma rays, X-rays, ultraviolet light, visible light, infrared radiation, and radio waves—was open to astronomy. New developments in technology, most of them from fields external to astronomy, have given astronomers new tools and ways of working. Engineers in the 1930s were the first to use radio astronomy, which received an enormous boost by the development of radar in World War II. X-ray astronomy and infrared astronomy began in the 1960s with the exploitation of detectors developed for nuclear and military applications.

Developments in space technology since the 1950s have influenced astronomy in two major ways. First, it became possible to place telescopes and other astronomical instruments on spacecraft orbiting above Earth's atmosphere. Since the atmosphere prevents most kinds of electromagnetic radiation from reaching the ground, this has freed astronomers to make use of the whole electromagnetic spectrum. Second, instrumented space probes have been sent to make direct investigations of the planets of our solar system, either from orbital reconnaissance or by landing on the surface. This has changed planetary astronomy from an observational backwater into the new and vibrant experimental discipline of planetary science.

But the need for large telescopes to gather light from faint and distant objects remains undiminished. The wider the lens or mirror of a telescope, the faster it can collect light and the fainter the objects it can observe. Back in 1900, the biggest telescope in the world was at Birr Castle, Ireland. Constructed half a century earlier by Irish politician and amateur astronomer William Parsons (Lord Rosse, 1800–1867), it had a metal mirror with a diameter of 1.8 m/6 ft and a tube that was 18 m/60 ft long supported between two castellated stone walls. It was moved up and down and from side to side by a system of winches and had a very restricted view of the sky.

Not until 1917 was a bigger and recognizably modern telescope constructed, the 2.5-m/100-in Hooker Telescope at Mount Wilson, near Pasadena, California. As we shall see, this powerful telescope played a crucial role in several decisive advances in the early decades of the twentieth century. Thirty years later, in 1948, the Hooker's big sister, the 5-m/200-in Hale Telescope on Mount Palomar, 70 km/43.5 mi north-northeast of San Diego, California, came into operation. With four times the collecting area of the Mount Wilson telescope, the Hale remains the most venerable, if no longer the largest, telescope in the world. Mention must be made here of the 6-m/20-ft Russian telescope at Zelenchukskaya, Russia, completed in 1976, although this telescope has had problems with its main mirror, and its performance has never matched the hopes of its designers.

A 100-inch Hooker telescope at Mount Wilson Observatory in California (1976). (Corbis/Denis di Cicco)

Also in 1948, a new kind of survey instrument, the Schmidt telescope, came into operation at Mount Palomar. Schmidts can photograph relatively large areas of sky, and the Palomar Observatory Sky Survey, completed in 1958, is an essential tool for every observatory. The complementary Southern Sky Survey, made by European Schmidts in Chile and Australia, was completed in the 1980s.

Since the 1960s, astronomers have chosen to place their new large telescopes at remote mountain sites far away from urban pollution and high enough to be above the clouds most of the time. Few countries can afford to build large telescopes on their own, and most of the new projects are international collaborations. The world's major new observatories are now on Mauna Kea in Hawaii, La Palma in the Canary Islands, and various sites in the Chilean Andes.

The Hubble Space Telescope (HST), a modest 2.4 m/94 in in diameter, owes its power to being permanently above the distorting effects of Earth's atmosphere that hinder ground-based telescopes. Launched by the United States in 1990, the HST was found to have a faulty primary mirror, and corrective optics were installed by a U.S. space shuttle crew in 1993. It has since worked flawlessly, returning spectacular images and streams of valuable data.

The Hubble Space Telescope in space. (NASA)

As the century drew to a close, several even bigger ground-based tele-scopes were becoming operational. The largest optical telescopes, the twin 10-m/33-ft Keck telescopes, which began work on Hawaii in the 1990s, are being rivaled by several in the 8-m/25-ft class, including a group of four at the European Southern Observatory's site at Cerro Paranal, Chile, that will act as a single telescope.

Radio Astronomy

The first significant advance outside the visible spectrum occurred in 1932, when Karl Jansky (1905–1950), a radio engineer at Bell Telephone Laboratories—an institution destined for a decisive contribution to cos-mology three decades later—in New Jersey, found that noise on transat-lantic radio-telephone links originated from the Milky Way. His discovery was taken up by fellow U.S. radio engineer Grote Reber (1911–), who built the first radio telescope, in his garden in Wheaton, Illinois, in 1937 and enjoyed the status of being the world's only radio astronomer until World War II.

As in many other aspects of life, the war changed everything for as-tronomers. The rapid development of radar led to technological innova-tion in the precise areas needed for radio astronomy. Indeed, the activities of Jansky and Reber notwithstanding, radio astronomy was independently rediscovered by defense scientists during the war. James Hey (1909–), a physicist working on radar for the British army, detected radar reflections from meteor trails high in the atmosphere and, in 1942, radio waves from the sun. After the war, he and others, with the benefit of huge amounts of war-surplus radar equipment, went on to set up university research groups to pursue radio astronomy, especially in Britain, Australia, the United States, and the Netherlands.

The most familiar type of radio telescope is the classic bowl-shaped antenna, the most significant of which was Manchester University's 76-m/250-ft Mark I radio telescope, built by British astronomer Bernard Lovell (1913–), that began operation at Jodrell Bank, Cheshire, England, in 1957. Within a few years, it was followed by similar large reflectors in Australia, the United States, and Germany. These enormous telescopes, far bigger than their optical counterparts, are superb for gathering up faint radio signals, but they are actually poorer than the unaided eye at resolving fine detail in the sky because of the much longer wavelength of radio waves.

An alternative approach was pioneered and developed in the 1940s at the Mullard Radio Astronomy Observatory, Cambridge University, England, under the direction of British radio astronomer Martin Ryle (1918–1984). Known as aperture synthesis, it involves linking several small telescopes together to give the resolving power of a much bigger instrument. The method was soon taken up all over the world; by the end of the twentieth century, important aperture-synthesis telescopes operated in numerous countries, including the United States, Britain, the Netherlands, and Australia. On an even larger scale, telescopes separated by thousands of kilometers can be made to operate as a single instrument, a technique known as very long baseline interferometry (VLBI). Some of the most spectacular images of astronomical objects are made with VLBI arrays spanning the entire world.

Other Astronomies

X-rays, like radio waves, were also discovered toward the end of the nineteenth century. Unlike radio waves, they cannot pass through Earth's atmosphere, so X-rays from space cannot be detected from the ground. Astronomers had reason to suppose that X-rays were emitted by the sun's extremely hot outer atmosphere, and in 1949 a team from the U.S. Naval Research Laboratory, Washington, D.C., succeeded in detecting solar X-rays by flying a detector on a captured German military rocket.

But the sun was bright in X-rays only because it was so close to Earth. No one expected that other stars would be significant X-ray sources. Nonetheless, in 1959, a private research company, American Science and Engineering, began a program of X-ray astronomy supported by a military contract to observe atmospheric nuclear explosions from high-flying sounding rockets. In 1962, while searching for X-rays created by bombardment of the moon by solar radiation, researchers found a bright source of X-rays in the constellation of Scorpius.

But rocket-borne instruments could operate for only a few minutes at most before returning to Earth. What was needed was an X-ray telescope mounted on a satellite orbiting permanently above the atmosphere. The first X-ray satellite, the U.S. spacecraft *Uhuru*, was launched in 1970 and detected more than three hundred X-ray sources in its twenty-eight-month survey of the whole sky. It was followed by others from the United States, Europe, and Japan that discovered many thousands of objects.

Even more energetic than X-rays are gamma rays, which are the subject of gamma-ray astronomy. The first cosmic sources of gamma rays were detected from balloons and satellites in the 1960s and 1970s. While many of them these sources correspond to known objects, others are still mysterious. Most puzzling of all are intense bursts of gamma rays that occur at random all over the sky at a rate of about one a day. First detected by U.S. military satellites in the 1960s, they remain enigmatic.

Ultraviolet astronomy, exploiting the broad band of wavelengths between visible light and X-rays, also suffers from atmospheric absorption over most of its range and developed in a similar way to X-ray astronomy. Following a partial satellite survey in 1968, the first dedicated ultraviolet (UV) satellite, the U.S.-European *International Ultraviolet Explorer,* was launched in 1978 and returned valuable data for eighteen years, a record for an astronomical satellite.

Infrared astronomy, bridging the gap between red-light and millimeter-radio waves, also benefited from detectors developed for military applications. Although infrared radiation from the sun had been detected by German-born British astronomer William Herschel (1738–1822) as long ago as 1800, astronomical observations are hampered by the absorbing effect of Earth's atmosphere at these wavelengths and by the large amounts of infrared radiation emitted by the atmosphere and, indeed, everything on Earth. It was not until the 1960s that the first survey of the infrared sky was conducted—by astronomers at the California Institute of Technology (Caltech) in Pasadena; it yielded more than fifty-six hundred sources, most of them cool stars.

Infrared astronomy developed rapidly in the 1970s, with new ground-based telescopes complemented by balloon-borne instruments. Infrared telescopes are similar in appearance to optical reflecting telescopes, except that the former are designed to prevent emission from the telescope itself reaching the detectors, which themselves are cooled to within a few degrees of absolute zero by liquid nitrogen and liquid helium. Near-infrared observations can be made from high mountain sites in a few relatively clear "windows" in the atmosphere, but for longer waves the only alternative was to mount telescopes on high-flying balloons or aircraft.

That all changed in 1983, when the U.S.-Netherlands-British *Infrared Astronomy Satellite* (IRAS) surveyed almost the whole sky at wavelengths inaccessible from the ground and detected a quarter-million sources. It was followed in 1995 by the European Space Agency's *Infrared Space Observatory* (ISO).

The final parts of the spectrum to be opened up are the submillimeter and millimeter waves, on the borderline between the far infrared and the microwave regions. Again, high mountain sites are required for the finely polished dishes of these telescopes.

The Rise of Planetary Science

So what have we learned about the universe in the twentieth century? Starting in our own backyard, the overall layout and constitution of the solar system was well established by 1900, with the exception that only eight planets were then known. Astronomers were still celebrating the discovery fifty-five years earlier of Neptune, whose existence had been predicted by its gravitational effects on Uranus—a triumph for Newtonian mechanics. Many hoped to find a ninth planet beyond Neptune by similar means, not least U.S. astronomer Percival Lowell (1855–1916), a wealthy U.S. amateur who began a search for the mysterious planet in 1905. Pluto was finally discovered in 1930 by U.S. astronomer Clyde Tombaugh (1906–1997) following a systematic photographic survey, and, at the close of the 1990s, it remained the farthest-known outpost of the sun's family of planets.

The planets are difficult objects to observe from Earth-bound telescopes, appearing as shimmering disks devoid of much detail. For the first half of the twentieth century, astronomers had little interest in the solar system, preferring to devote their attentions to the more rewarding study of stars and galaxies.

In 1957, the then–Soviet Union launched *Sputnik 1*, the first artificial satellite of Earth. Though the motivation behind both the Soviet and the U.S. space programs was primarily national prestige at the height of the cold war, astronomers and geologists were quick to see the potential for science of being able to send packages of instruments through space to investigate objects in the solar system at close quarters. Within two years, a Soviet spacecraft had crash-landed on the moon, and another had passed behind the moon, transmitting images of the unseen far side. The latter, in particular, proved that automatic space probes could achieve feats that were impossible from Earth.

The moon, being by far the nearest body to Earth, received special attention. For reasons that had little to do with science, the United States and the Soviet Union sent no fewer than seventy-eight space missions toward the moon between 1958 and 1974. A flotilla of automatic probes crashed on the moon, orbited around it, and, eventually, with Soviet probe *Luna 9* in 1966, landed safely and transmitted pictures from the surface. In 1969, two U.S. astronauts, Neil Armstrong (1930–) and Edwin ("Buzz") Aldrin (1930–), from *Apollo 11* landed in the Sea of Tranquility and returned home with samples of rocks and dust. Over the next three and a

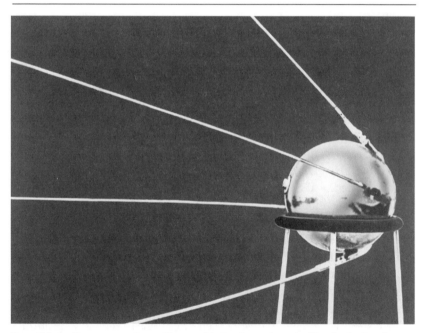

Sputnik I, the first man-made satellite to orbit Earth, was launched by the Soviets from an undisclosed location on October 4, 1957. (Corbis-Bettmann)

half years, a further five *Apollo* landings were to follow, and, although only the final one, *Apollo 17* in 1972, included a geologist, a wealth of scientific information was obtained. The picture we now have of the moon is of an essentially dead world, formed by the collision of a Mars-size object with Earth in the early years of the solar system.

Unlike the moon, the former superpowers never had any serious strategic interest in the other planets, and the motive for exploring them has been almost entirely scientific. Our next-nearest neighbor in space is the planet Venus. In 1900, the chief importance of Venus was that it occasionally passed across the face of the sun, allowing astronomers to calculate its precise orbit and, from there, the scale of the solar system, but little else was known. Shrouded in impenetrable cloud, Venus reveals few of its secrets to Earth-bound telescopes. Roughly the same size and mass as Earth, it had been the subject of much speculation. As late as the 1960s, some astronomers still fancied that Venus could be covered with oceans teeming with life.

It was not until the Soviet probe *Venera 4* entered that planet's atmosphere in 1967 that scientists discovered that Venus possessed a crushing atmosphere of carbon dioxide ninety times the pressure on Earth and temperatures up to a searing 480°C/900°F. Further probes, including the landing of *Venera 7* in 1970, confirmed that Venus is the victim of a runaway greenhouse effect in which any water it once possessed has been lost. Several probes have mapped its surface by radar, an example of military technology successfully exploited for space science. The very productive U.S. Magellan Project—the *Magellan* space probe was launched in 1989—has shown Venus to be covered in lava flows and numerous volcanoes, confirming that the planet has developed very differently than Earth.

Mars also proved to be a disappointment to those expecting Earth-like conditions. As the twentieth century began, U.S. astronomer Percival Lowell claimed to have seen no less than 160 canals crisscrossing the Martian landscape, which he interpreted as an irrigation network constructed by an intelligent civilization. As the century wore on, belief in the canals among scientists waned, but many retained a residual hope that Mars might be the one planet in the solar system harboring some kind of life. Reliable reports of seasonal color changes across the planet kept hopes alive that, even if there were not canal builders on Mars, there might be simple plant life.

The U.S. probe *Mariner 4* dashed all such hopes when, flying past the planet in 1965, it showed Mars to be heavily cratered like the moon with no traces of canals. In 1976, the two U.S. *Viking* missions dropped landers on to the surface and surveyed the planet from orbit. Mars has majestic volcanoes (though no longer active), rolling plains, a canyon thousands of kilometers long and, intriguingly, what appear to be dried-up river-beds. If Mars is dry now—the polar caps are carbon-dioxide frost, not water ice—it certainly had flowing water in the distant past.

Interest in Mars was renewed in 1996, when Daniel Goldin of the National Aeronautics and Space Administration (NASA) announced that a meteorite found in Antarctica, and believed to have come from Mars, contained evidence of primitive life, including what appeared to be fossilized bacteria. Although the evidence is not now regarded as compelling, the discovery gave a huge boost to NASA's plans for a new series of Mars probes, starting with the *Mars Pathfinder*, which was launched in late 1996 and landed a small roving vehicle on the planet in 1997.

The innermost planet, Mercury, is so small and close to the sun that hardly anything new was discovered about it in the first sixty years of the twentieth century. Its chief claim to fame was that its motions could not be entirely explained until Einstein's theory of general relativity was published in 1915. As a result of an erroneous observation by Italian astronomer Giovanni Schiaparelli (1835–1910) in the 1880s, it had long been believed that Mercury always kept the same face toward the sun, resulting in one baking hemisphere and one freezing one. It was not until the 1960s that this error was corrected, when radar observations from Earth proved that the planet rotated once every fifty-nine days. Almost everything else we know about Mercury was gained in 1974, when the sole spacecraft to visit the planet, the U.S. probe *Mariner 10,* sailed past, sending back pictures. Mercury, it turned out, was as barren and cratered as the moon and lacked an atmosphere. No spacecraft has visited Mercury since, and, as the twentieth century came to a close, no visits had been planned.

The giant planets Jupiter and Saturn, on the other hand, present much more rewarding views to Earth-bound telescopes. Even in 1900, it was clear that, unlike the inner planets, Jupiter was a gaseous world wrenched by perpetual storms, perhaps still condensing from a primordial nebula. The four main moons had been known since the time of Italian astronomer and physicist Galileo Galilei (1564–1642). Saturn, too, was largely gaseous, and logic already dictated that the distinctive rings surrounding the planet could not be solid but were likely to consist of swarms of small objects.

By the 1960s, telescopic observations had demonstrated that both planets consisted largely of hydrogen and helium, with traces of ammonia and methane—very different from the so-called terrestrial planets like Earth. The first spacecraft to fly past Jupiter and Saturn—the U.S. *Pio-*

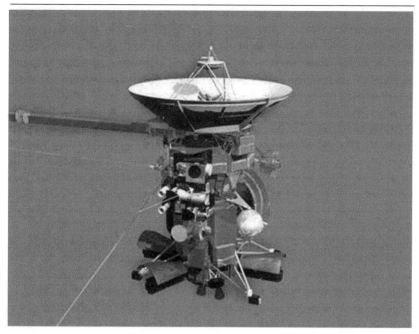

A diagram of the *Cassini* spacecraft. (NASA)

neers 10 and *11* in the mid-1970s—were followed by the very successful U.S. *Voyager* probes (1979–1981) and later by the U.S. probe *Galileo*, launched in 1989, which went into orbit around Jupiter and dropped a probe into the atmosphere in 1995. An international mission to Saturn, the *Cassini* spacecraft, left Earth in 1997 and was to land a probe on Saturn's largest moon, Titan, in 2004.

Voyager 2 went on to fly past Uranus in 1986 and Neptune in 1989, leaving Pluto as the only planet so far unvisited from Earth.

In 1910, the world was dazzled by the close approach of Halley's comet to Earth. Upon its return in 1986, Halley was met by a flotilla of spacecraft, including the European probe *Giotto* that took close-up pictures of the comet's icy nucleus. Comets are now believed to be remnants of the primeval material out of which the planets formed. Also widely accepted is the idea put forward by Dutch astronomer Jan Hendrik Oort (1900–1992) in 1950 that comets are visitors from a vast cloud—the Oort cloud—of perhaps a thousand million comets inhabiting the farthest reaches of the solar system.

As for the origin of the solar system, astronomers have returned to a version of the nebular hypothesis of French mathematician Pierre Laplace

(1749–1827) that the sun and its retinue of planets condensed out of a cloud of dust and gas. That notion fell into disfavor during the early decades of the twentieth century in the wake of an alternative idea known as the collision theory. It held that the planets were formed by the condensation of debris resulting from a near-collision between the sun and a passing star. Many a popular book showed a cigar-shaped cloud being drawn from the sun and coagulating to form the planets. The nebular hypothesis predicts that planetary systems should be commonplace (the exact opposite of the collision theory), so it was particularly gratifying that the 1990s should see the first discoveries of planets in orbit around nearby stars.

In a real sense, the twentieth century has seen both the maturity of planetary astronomy and its growth into a new and distinct discipline of planetary science. Planetary scientists may be astronomers, but they are more likely to be geologists. The planets have become worlds.

The Sun and the Stars

By the end of the nineteenth century, astronomers knew full well that the stars were suns and that the sun was a star, though their source of energy, which was of crucial importance, was mysterious. The biblical notion that the world was a mere six thousand years old had long been rejected by science, and it was already clear that the sun could not be powered by any kind of chemical reaction—the sun was not "burning" in any conventional sense of the word—since such reactions could have sustained the sun's output at its present rate for only a few thousand years at most. By the last decade of the nineteenth century, physicists were confident that the source of the sun's energy was gravitational: The sun was slowly contracting under its own weight, and the energy released was sufficient to keep the sun shining for tens of millions of years.

Geologists, on the other hand, had concluded that hundreds of millions or even billions of years were required for the deposition of rocks and other geological processes, and the sun could obviously be no younger than Earth. This view gradually gained ground as evidence from the new theory of evolution pointed to an ancient Earth, and, as the century closed, there seemed to be no way of reconciling these deeply conflicting, yet equally rational, views.

In the meantime, progress was being made on other fronts. One of the most powerful tools in astronomy—the spectroscope—had been refined by many scientists since the early nineteenth century. Spectroscopy allowed astronomers to split up the light from stars into its component colors and identify the tell-tale signatures of individual chemical elements. By this means, astronomers had begun to learn the compositions of stars and to classify them into different types. They were able to measure their speed of motion through the Doppler shift, in which movement along the line of sight causes measurable changes in the wavelength of emitted light: a lengthening (red shift) for objects moving away from the observer and a shortening (blue shift) for objects moving toward the observer. By such means, studies of binary-star systems had even yielded the masses of some stars, which turned out to be not too dissimilar from that of the sun.

In the fifty years since cameras were first turned to the sky, photography had become a powerful tool in astronomy and had already led to many important discoveries, as well as the provision of reliable catalogs and atlases. Long exposures allowed astronomers to see objects and detail that could not be discerned by the eye, even with large telescopes. Photographs made possible the precise measurement of stellar positions and, with that, came the discovery that, far from being fixed, as traditionally supposed, some stars were seen to be slowly moving across the sky.

This, in turn, led to attempts to determine the distances of stars from the sun by measuring their apparent positions at different times of year, a method known as parallax. By the turn of the twentieth century, several dozen distances had been measured, and it was clear that the nearer stars were, in fact, at immense distances from the sun, such that light would take several years to make the journey. Clearly, the stars were of similar intrinsic luminosity to the sun.

In the first decade of the new century, much work was done on classifying stars according to their spectra, which indicated their surface temperature. Two astronomers working independently, Ejnar Hertzsprung (1873–1967) in Denmark and Henry Russell (1877–1957) in the United States, discovered that stars could be classified also by their luminosity. When stars were plotted on a graph of luminosity against surface temperature, the bulk of stars fell into a group known as main-sequence stars, and the remainder into a group known as giants. It soon became clear that these two groups represented different evolutionary stages in the life of a star. The Hertzsprung-Russell diagram, as the graph became known, became and remains the key to understanding stellar evolution.

In the 1920s, British astrophysicist Arthur Eddington (1882–1944) and others attempted to solve the solar-energy problem by proposing that the source of the sun's energy is the nuclear fusion of hydrogen to helium. Considering that little was yet known about the nucleus of the atom, this was a bold supposition, and progress had to await the further development of nuclear physics. In 1938, German-born U.S. astronomer Hans Bethe (1906–) and German physicist Carl von Weizsäcker (1912–) independently calculated that the sun and all other main-sequence stars were powered by a chain of reactions that resulted in the fusion of four hydrogen nuclei to create a nucleus of helium. Most gratifying, the sun must have enough hydrogen fuel to last ten billion years, thus bringing the age of the sun at last into line with geological evidence of the age of Earth.

From this discovery, the way was open for others to further develop the theory of stellar evolution that accounted for all of the main features of the Hertzsprung-Russell diagram. The advent of powerful computers in the 1960s accelerated this process, allowing ever more sophisticated models to be applied. The general idea was that stars are formed by the gravitational contraction of interstellar gas clouds, followed by a long period of hydrogen burning on the main sequence. When the hydrogen runs out, the star contracts further, heating up until the helium can be fused to create heavier elements. At this point, the star has become a red giant, the second main group in the Hertzsprung-Russell diagram. Depending on its mass, the star may go through several such cycles, each time burning heavier elements at higher temperature, until all of the fuel is used up. But what then?

Low-mass stars were already known to end up as Earth-size objects known as white dwarfs, which are simply the cooling, shrunken embers of giant stars that have run out of fuel. As long ago as 1930, Indian-born U.S. astrophysicist Subrahmanyan Chandrasekhar (1910–1995) calculated that white dwarfs could not have a mass greater than 1.4 times the mass of the sun, since a heavier star could not support its own weight after nuclear reactions had died away. But in the 1930s, no one could say for certain what would happen to a star that exceeded the Chandrasekhar limit.

The answer came in 1967 when Irish radio astronomer Jocelyn Bell Burnett (1943–) and British astronomer Antony Hewish (1924–) at Cambridge University, England, discovered the first fundamentally new type of star: pulsars. Pulsars, they found, were sources of pulsing radio waves that turned out to be extremely compact and dense stars only a few kilometers across, spinning rapidly on their axes. Theorists quickly invoked a 1932 prediction by Soviet physicist Lev Landau (1908–1968) and showed that pulsars were so-called neutron stars, the collapsed cores of massive stars that had finally run out of nuclear fuel and exploded as supernovae. Here was another end point of stellar evolution: Stars too massive to become white dwarfs ended their lives in a supernova explosion, leaving behind a neutron star.

Astronomers were astonished when a supernova exploded in a nearby companion galaxy to the Milky Way in 1987, the first to be visible to the unaided eye since 1604. A wealth of information was gained about the physics of supernova explosions, though the expected pulsar was not detected.

But the strangest star of all is also the most elusive. Neutron stars, too, have a maximum mass, perhaps two to three times the mass of the sun. Beyond that, they will collapse and be crushed out of existence in a black hole. A black hole is an object whose gravitational field is so strong that even light cannot escape from it. The sun, for example, would disappear into a black hole if it were crushed to a radius of 3 cm/1.2 in. Such objects were predicted in 1916 by German astronomer Karl Schwarzschild (1873–1916), as a consequence of Einstein's theory of general relativity, but were regarded as theoretical curiosities until the 1960s.

When astrophysicists tried to explain the wholly unexpected X-ray "stars" discovered from 1962 onward, black holes provided a powerful source of energy. A normal star in a close orbit around a black hole or a neutron star would be subject to intense gravitational forces that would drag gas from its surface down toward the compact object. The falling, swirling gas would be heated to the point at which it emitted X-rays.

Unlike neutron stars, black holes have been extraordinarily hard to confirm, since they can be detected only by their gravitational effects on nearby matter. Few astrophysicists doubt that they exist, yet, despite several prime suspects, conclusive identifications remain lacking. It turns out that the best evidence for black holes is to be found not among the stars of our own galaxy, but far out in space in the hearts of some of the most violent objects in the known universe.

Galaxies

In 1900, astronomers knew that the sun was part of a larger system of stars known as the Milky Way, or galaxy, which was regarded as a roughly lens-shaped structure containing innumerable stars and misty nebulae. But, apart from the distances of a few dozen nearby stars, astronomers had no idea of the size of the Milky Way or, indeed, of the universe itself.

The mysterious nebulae—several thousand of them—had been cataloged all over the sky since the days of William and Caroline Herschel's systematic search for them in the late eighteenth century. These appeared to the eye as faint shining clouds, often punctuated with numerous stars. Some were irregular in shape; others had curious spiral forms. Spectroscopic analysis had revealed that the Great Nebula in Orion contained hot hydrogen gas, but others, like the Great Nebula in Andromeda, did

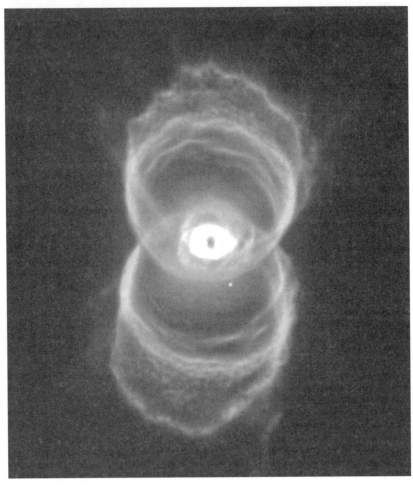

A photograph of an hourglass nebulae (1996). (NASA)

not. What were they? Some astronomers speculated that nebulae may lie hundreds of light-years from the sun, but they generally despaired of ever being able to measure a distance to them. In 1900, hundreds of light-years seemed an appalling distance, surely approaching the limits of the universe itself.

A major advance came in 1912, when U.S. astronomer Henrietta Swan Leavitt (1868–1921) published her investigations of a class of stars known as Cepheids, whose light output changes in a regular fashion. She found that the Cepheids varied over a period directly related to their luminosity: The slower the variation, the more luminous the star. If one could measure the period of distant Cepheids, a simple task, from their apparent brightness one could then calculate how far away they were.

In 1918, U.S. astronomer Harlow Shapley (1885–1972), at the Mount Wilson Observatory in California, used a similar method to measure the distance to dense agglomerations of stars known as globular clusters. He noticed that the clusters seemed concentrated in one part of the sky and surmised that they were grouped in a roughly spherical shape around the distant center of our galaxy. He was able to calculate that the center of the galaxy was some 45,000 light-years distant. In fact, Shapley's estimate was too high by about a factor of two, but only because he was unaware that the space between the stars was not empty but was scattered with dust, making the clusters look dimmer and, so, more distant than they really were. Discovered in 1930 by Swiss-born U.S. astronomer Robert Trumpler (1886–1956), interstellar dust explained why we appear to be at the center of our galaxy—we simply cannot see the dense, but obscured, central regions.

By the early 1920s, then, the universe had grown considerably. Our galaxy was, indeed, lens shaped, but with the sun far from the center. Measurements of Cepheids in two misty patches called the Magellanic Clouds—now known to be small galaxies in their own right—had placed them at distances of a hundred thousand light-years, just outside the confines of the Milky Way. Was this the extent of the universe, or was there more outside? By 1920, views had crystallized into two opposing camps: One held that the nebulae were relatively small objects within the Milky Way; the other argued that they were, in fact, assemblies of stars, like the Milky Way itself, at vast distances from us.

In 1924, U.S. astronomer Edwin Hubble (1889–1953) reported that he had detected a number of Cepheids in the Andromeda nebula. The nature of the Great Nebula in Andromeda, with its exquisite spiral shape, had long been a mystery, but Hubble, using the new 2.5-m/100-in Hooker Telescope on Mount Wilson, had not only resolved the mists into stars but had identified Cepheids among them. Using the now routine period-luminosity method established by Leavitt, he calculated that the nebula was no less than 930,000 light-years from the sun and, therefore, far beyond the bounds of the Milky Way.

Many other "nebulae" proved to be galaxies like our own and even more distant than Andromeda. The universe was now larger than had ever been imagined and populated by majestic "island universes" of galaxies, as originally proposed by German philosopher Immanuel Kant (1724–1804) as far back as 1755. Hubble went on to devise a classification scheme of galaxies based on the now familiar types of spirals and ellipticals.

During the enforced blackouts of World War II, German-born U.S. astronomer Walter Baade (1893–1960) used the 2.5-m/100-in Mount Wilson telescope to examine the stars in the Andromeda galaxy. He discovered that stars in the central bulge of the galaxy were different in nature from those farther out. In particular, the Cepheids in the two populations were different and had a different period-luminosity relationship. When he allowed for this difference, he found that Andromeda was twice as distant as Hubble had calculated, as were all of the other galaxies. The universe was even bigger than had been supposed.

With greater understanding of stellar populations and the effects of obscuring dust, astronomers were making great progress in working out the structure and dynamics of our own galaxy. A great leap came in 1951, when U.S. radio astronomers found that they could detect radio emission at a wavelength of 21 cm/8.3 in from vast clouds of hydrogen right across the galaxy, a phenomenon that had been predicted by Dutch astronomer Hendrik van de Hulst (1918–) seven years earlier. By mapping the positions of the clouds, radio astronomers constructed the first large-scale maps of the galaxy, showing that it had a spiral form just like other galaxies. By the 1960s, astronomers had developed a plausible theory to explain the spiral pattern in terms of waves of enhanced density.

Meanwhile, radio astronomers were finding intense sources of radio waves in the sky, a few of which, like a powerful object discovered by British physicist James Hey in 1946, seemed to coincide with known galaxies. Why were these galaxies emitting radio waves while others were not? In 1963, Australian and U.S. astronomers identified a radio source with a faint starlike object possessing a peculiar spectrum. Dutch-born U.S. astronomer Maarten Schmidt (1929–) showed that the spectrum made sense if it was moving away from Earth at an astonishing 47,400 km per second (kps)/28,400 miles per second (mps), or some 16 percent of the speed of light. According to the "red shift" law discovered by Hubble (see next section below), the object was no less than three billion light-years away. This proved to be a quasar, the first of thousands to be discovered, most of them at even greater distances.

If the red shift is, indeed, a correct measure of distance, quasars emit a hundred times the luminosity of a whole galaxy from an object no bigger than our solar system. Gradually, astronomers have come to understand that quasars are sited in the cores of distant galaxies. Their source of energy is believed to be a supermassive black hole—perhaps a hundred million times the mass of the sun—into which material is falling, releasing vast amounts of gravitational energy. On this principle, theorists have been able to present a plausible model of a whole range of "active" galaxies, including quasars, radio galaxies, and other similarly energetic objects. Some speculate that a smaller black hole may even be lurking in the core of our own galaxy.

Cosmology

In 1900, astronomers had conflicting notions about the age of Earth and the sun, but science had little to say yet about the age of the universe as a whole or how it had come into being. The twentieth century would see rapid progress in this area, with the question essentially settled by its end.

In 1915, German-born U.S. physicist Albert Einstein (1879–1955) published his theory of general relativity. Ten years earlier, his theory of special relativity had said startling things about the nature of time and

During a lecture delivered by Dr. Gustave Stromberg in Pasadena, Albert Einstein writes some chalk figures on the blackboard thoroughly confusing everyone, including Professor Stromberg (17 January 1931). (Corbis-Bettmann)

space, but it had been restricted to objects moving at constant speed with respect to each other. Now Einstein tackled the question of accelerating objects and showed how gravitational forces could be understood in terms of local distortions of time and space. He cleared up the long-standing problem of the orbit of Mercury and predicted that light would be bent on passing close to the sun, a prediction dramatically verified during the total solar eclipse of 1919.

Even more remarkable, general relativity now permitted theorists to devise models that represented the universe as a whole. There were many such models, but one thing they all had in common was expansion or contraction. A static universe was ruled out, much to the consternation of Einstein, who could not believe that the universe might be expanding and added a fudge factor—the "cosmological constant"—to his equations to oppose gravity and keep it static, a move he later regretted.

From 1912 onward, U.S. astronomer Vesto Slipher (1875–1969) began to measure the speed of motion of spiral nebulae by means of the Doppler effect. By measuring the precise amount of shift, Slipher could calculate the speed. Slipher found that almost all of the spiral nebulae had red shifts and so appeared to be moving away from the sun, often at very high speeds.

In the mid-1920s, Edwin Hubble, who had first recognized the true nature of the spiral nebulae, embarked with Milton Humason (1891–1972) on a series of precise speed measurements with the 2.5-m/100-in Mount Wilson telescope. In 1929, he announced that the apparent recession speed of a galaxy was directly proportional to its distance. It was immediately apparent that this "Hubble's Law" was consistent with the kind of universal expansion that the cosmologists had been considering. Moreover, if the expansion were projected backward, much as a film can be run backward, all of the galaxies would have been crowded together in a small volume at some definite instant in the past. Thus was born the notion that the universe is of finite age and came into being in a violet event—later dubbed the Big Bang—from which it has been expanding ever since.

From the 1940s onward, an alternative theory began to gain ground. It held that the expansion could be accounted for if matter were continuously being created to maintain the density of the universe. Thus, there was no beginning; the universe had existed forever. The architects of this steady-state theory—Austrian-born British mathematician and cosmologist Hermann Bondi (1919–), Austrian-born U.S. astronomer Tommy Gold (1920–), and British astrophysicist Fred Hoyle (1915–)—pointed out that the continuous creation of matter was no more startling than the initial creation of matter demanded by the favored Big Bang model. The stage was set for a contest between the Big Bang and steady-state theories that engaged public interest throughout the 1950s. Although surveys of distant galaxies by radio astronomers were seeming to favor the Big Bang interpretation, by the early 1960s the confidence of earlier decades seemed to be replaced with confusion.

In 1965, two scientists working at Bell Telephone Laboratories in New Jersey, German-born U.S. radio engineer Arno Penzias (1933–) and U.S. radio astronomer Robert Woodrow Wilson (1936–), were attempting to trace the source of background noise in an antenna used for tracking communications satellites. The noise appeared as a steady hiss detected from all parts of the sky. Its significance was recognized by U.S. physicist Robert Dicke (1916–1997), who had realized that the universe should be bathed in radiation from the Big Bang and that this so-called cosmic background radiation should be detectable with radio telescopes, a suggestion originally made by Russian-born U.S. physicist and cosmologist George Gamow (1904–1968) and colleagues in the 1940s. Moreover, the intensity of the radiation detected by Penzias and Wilson was close to that predicted by the Big Bang model. From then on, the Big Bang theory swept the board, with the steady-state alternative all but abandoned, despite modifications to account for the cosmic background radiation.

But not everyone was willing to take the leap of faith that the red shifts of distant galaxies and quasars were cosmological. Proponents of the steady-state view, in particular, drew attention to many instances of two or more galaxies or quasars of differing red shift that appeared to be physically associated with each other and so, presumably, at the same distance from Earth. Their opponents put these alignments down to chance, and the "red-shift controversy" simmered away through the 1970s. At the end of the twentieth century, it had yet to be entirely settled. Nonetheless, the Big Bang theory has become widely accepted as the best working model of the universe.

Will the universe continue expanding indefinitely? Or will there come a time when the expansion halts and the universe starts to contract? It seems unlikely we could ever find out, but, remarkably, all we need to know is the average density of the universe. If the density is less than a certain critical density, the expansion will go on forever; if greater, the expansion will eventually reverse. It seems that, whatever the density is, it is so close to the critical density that it must have been unimaginably close in the early moments of the universe. Coincidence?

Quite why the density seems to have this very special value was accounted for in the early 1980s by the concept of inflation, devised by U.S. astrophysicist Alan Guth. In this scenario, the universe undergoes an extremely rapid expansion in the first fraction of a second after the Big Bang before settling down to the normal "Hubble" expansion. It is this early explosive ballooning that ensures that the subsequent density will always be close to critical. As a bonus, inflation also explains why the cosmic background radiation has the same brightness in all directions— all of the matter in the visible universe had attained the same temperature even before inflation started.

Nonetheless, we still do not know the density of the universe precisely enough to say with any confidence what will happen in the future. One problem is that there is more material in the universe than is apparent at first sight. Since the 1930s, astronomers have suspected that the visible matter—stars and galaxies—is supplemented by unseen dark matter detectable only by its gravitational effects on other objects. Dark matter has been detected within galaxies, in the outer halos of galaxies, and inside clusters of galaxies. By some reckonings, dark matter makes up more than 95 percent of the mass of the universe, and so plays a dominant role in determining the future course of the universe. What is more, only a small fraction of the dark matter can be in the form of conventional objects, such as burned-out stars. Most of it must be an exotic kind of matter that has not yet been detected. At the end of the twentieth century, the nature of dark matter was one of the most active areas of observational cosmology.

Cosmology benefited little from the development of space science until 1989, when the United States launched a spacecraft, the *Cosmic Background Explorer* (COBE), into Earth orbit to make precise measurements of the cosmic background radiation. Almost immediately, COBE scientists determined that the radiation was exactly of the form expected by the Big Bang model and of a temperature of 2.73K (–270.43°C/–454/76°F), or 2.73° above absolute zero. In 1992, COBE data resulted in a second big discovery: The background radiation varied in brightness from place to place on the level of one part in a hundred thousand. These tiny variations, or ripples, are interpreted as evidence of clusters of galaxies forming in the early life of the universe. Though hailed in the press as a major discovery, such structure was widely predicted from models of galaxy formation, and astronomers would have been very concerned if COBE had not seen it.

Questions remain, though. Much to their embarrassment, astronomers still do not know very well how fast the universe is expanding and, hence, its age. These are very difficult observations to make, and different methods give different answers. The Hubble constant, which measures the rate of expansion, is known only to about 20 percent at best. Given the already large uncertainty in the density of the universe, the best guess of its age is nine billion to fifteen billion years.

Yet, the history of science warns us that, as soon as we think we understand the universe, nature pitches a surprise. By the late 1990s, astronomers had learned how to detect supernova explosions in large numbers of extremely distant galaxies. To their surprise, these supernovae appeared fainter than expected and, so, at a greater distance. This apparently innocuous discovery forced cosmologists to contemplate the uncomfortable possibility that the expansion of the universe was actually speeding up.

Until then, it had been assumed that the expansion was slowing down due to the gravitational pull of the galaxies on each other. But, if the expansion is accelerating, then some new force must be driving the expansion, a force arising from the fabric of space itself. Is it possible, cosmologists asked, that Einstein was, in fact, correct when he added his cosmological constant to the equations of general relativity to oppose the effects of gravity?

At the close of the twentieth century, cosmology was once again in turmoil.

—A. W. Jones

2

Chronology

1903

Russian scientist Konstantin Eduardovich Tsiolkovsky writes "Investigations of Space by Means of Rockets," in which he outlines the use of liquid-propelled rockets to escape Earth's gravity.

1905

U.S. astronomer Percival Lowell, after a study of the gravitation of Uranus, predicts the existence of the planet Pluto.

1905–1907

Danish astronomer Ejnar Hertzsprung discovers that there is a relationship between the color and the absolute brightness of stars and classifies them according to this relationship. The relationship is used to determine the distances of stars and forms the basis of theories of stellar evolution.

1906

German astronomer Max Wolf discovers the asteroid Achilles. It is the first of approximately one thousand asteroids known as Trojan planets that form an equilateral triangle with Jupiter and the sun. An example of the solution to the three-body problem in astronomy (that is, the determination of the motion of three bodies under the influence of their mutual attraction), it revolves around the sun in the Lagrangian point of Jupiter's orbit.

U.S. astronomer Percival Lowell publishes *Mars and Its Canals,* in which he argues that the canal-like markings on Mars are irrigation canals built by intelligent creatures.

1908

U.S. astronomer George Ellery Hale discovers that sunspots have magnetic fields.

1909

British astronomer John Evershed discovers that gases radiate from the centers of sunspots.

1910

May 19. Halley's comet—which comes near Earth roughly every seventy-five years—returns, with Earth passing through the comet's tail. In the United States, it is regarded by some as announcing the end of the world. "Comet Pills," allegedly an antidote to the poisonous gases thought to be in the comet's tail, also sell well.

1911

German-born U.S. physicist Albert Einstein calculates the deflection of light caused by the sun's gravitational field.

1912

U.S. astronomer Edward Barnard discovers Barnard's Star. Second-nearest star to the sun, it displays the greatest movement of any star relative to others.

U.S. astronomer Henrietta Leavitt establishes a relationship between the period and the luminosity of Cepheid-variable stars (stars that pulsate and vary regularly in brightness). The relationship is later used to calculate interstellar and intergalactic distances.

1912–1925

U.S. astronomer Vesto Slipher measures the radial velocities of spiral nebulae by examining small changes in the Doppler effect that suggest that they must be external to our galaxy. His work is later built on by U.S. astronomer Edwin Hubble to show that the universe is expanding.

1913

Danish astronomer Ejnar Hertzsprung introduces a luminosity scale of Cepheid-variable stars to measure their distance.

U.S. astronomer Henry Russell shows that there is a correlation between a star's brightness and its spectrum. The correlation is important in determining stellar distances.

1914

British astronomer John Franklin publishes the *Franklin-Adams Charts,* the first photographic star charts of the entire sky.

British astrophysicist Arthur Eddington publishes *Stellar Movement and the Structure of the Universe,* in which he theorizes that spiral nebulae are galaxies similar to the Milky Way.

1915

Proxima Centauri is discovered. The closest star to the sun, it is 4.2 light-years away from Earth.

1916

German astronomer Karl Schwarzschild offers a solution to Einstein's gravitational-field equations that predicts the existence of black holes.

1917

The NASA Langley Research Center is established in Hampton, Virginia. It quickly becomes the world's leading aeronautical research center.

The 2.5-m/100-in Hooker reflecting telescope is installed at Mount Wilson Observatory, near Los Angeles, California. It is the world's largest reflecting telescope to date.

Dutch astronomer Willem de Sitter shows that Einstein's theory of general relativity implies that the universe must be expanding.

Ariel view of the Langley Field Research Center in Hampton, Virginia. (NASA)

U.S. astronomer Harlow Shapley determines that the sun is situated some 30,000 light-years from the central plane of the galaxy.

1918

U.S. astronomer Harlow Shapley estimates the size of the Milky Way as 45,000 light years.

1919

The International Astronomical Union (IAU) is founded to promote international cooperation in astronomy.

U.S. astronomer William Pickering predicts the existence and location of the planet Pluto.

U.S. inventor Robert Hutchings Goddard publishes "A Method of Reaching Extreme Altitudes," which outlines the use of rockets as a means to reach the moon.

1920

German-born U.S. astronomer Walter Baade discovers the asteroid Hidalgo, which is unusual in that its orbit is tilted out of the plane of the solar system by 43°.

U.S. physicist Albert Michelson, using a stellar interferometer, measures the diameter of the star Betelgeuse to be 386.1 million km/241.4 million mi, which is about three hundred times the diameter of the sun. It is the first time an accurate measurement of the size of a star other than the sun has been made. (The diameter of Betelgeuse is now thought to be 1,100 million km/700 million mi, about eight hundred times the diameter of the sun.)

1922

May 12. An 18-metric-ton/20-ton meteorite lands in a field near Blackstone, Virginia, leaving a 46 m²/500 ft² hole in the ground.

1923

German mathematician Hermann Oberth publishes *Die Rakete zu den Planetenräumen [The Rocket into Interplanetary Space]*, a treatise on spaceflight in which he is the first to provide the mathematics of how to achieve escape velocity.

German-born British physicist Frederick Lindemann investigates the size of meteors and the temperature of the upper atmosphere.

1924

U.S. astronomer Edwin Hubble demonstrates that certain Cepheid-variable stars are several hundred thousand light-years away from Earth and, thus, outside the Milky Way galaxy. The nebulae in which they are found are the first galaxies to be discovered that are proved to be independent of the Milky Way.

1926

British astrophysicist Arthur Eddington publishes *Internal Constitution of the Stars*, in which he shows that the luminosity of a star is a function of its mass.

Swedish astronomer Bertil Lindblad discovers that the Milky Way rotates around its center. One rotation takes 210 million years.

U.S. astronomer Heber Curtis and Swedish astronomer Knut Lundmark contend that spiral nebulae are galaxies similar to the Milky Way.

U.S. inventor Robert Hutchings Goddard conducts a static test of a liquid-propelled rocket.

March 26. U.S. inventor Robert Hutchings Goddard achieves the first flight of a liquid-propelled rocket, at Auburn, Michigan. It reaches an altitude of 12 m/41 ft.

1927

Belgian astronomer Georges Lemaître proposes that the universe was created by an explosion of energy and matter from a primeval atom—the beginning of the Big Bang theory.

U.S. astronomer Edwin Hubble shows that galaxies are receding and that the farther away they are, the faster they are receding.

The Verein für Raumschiffahrt (Society for Space Travel) is founded in Germany for rocket experimentation.

1928

Astronomers Joseph Moore and Donald Menzel discover that Neptune's moon Triton rotates in the retrograde direction—that is, opposite that of Neptune's spin.

British physicist James Hopwood Jeans proposes the steady-state hypothesis, which states that the universe is constantly expanding and maintaining a constant average density through the continuous creation of new matter.

1929

British physicist James Hopwood Jeans publishes *The Universe around Us,* which helps popularize astronomy.

Russian-born U.S. cosmologist George Gamow, U.S. physicist R. Atkinson, and German physicist Fritz Houtermans suggest that thermonuclear processes are the source of solar energy.

German mathematician Hermann Oberth publishes *Wege zur Raumschiffahrt [Ways to Spaceflight],* which discusses ion rockets and electric propulsion.

U.S. astronomer Edwin Hubble publishes Hubble's Law, which states that the ratio of the speed of a galaxy to its distance from Earth is a constant (now known as Hubble's constant).

U.S. astronomer Henry Russell publishes "Stellar Evolution," in which he suggests that stars begin as huge cool red bodies, shrink to become hot yellow stars and then hot white and blue dwarfs, and then shrink to become cool red stars.

1930

February 18. U.S. astronomer Clyde Tombaugh, at the Lowell Observatory, Flagstaff, Arizona, discovers the ninth planet, Pluto.

Swiss-born U.S. astronomer Robert Trumpler discovers the existence of interstellar material that reduces the apparent brightness of distant stars.

1932

U.S. radio engineer Karl Jansky discovers that the interference in telephone communications is caused by radio emissions from the Milky Way. He thus begins the development of radio astronomy.

U.S. scientist Carl David Anderson, while analyzing cosmic rays, discovers positive electrons (positrons), the first form of antimatter to be discovered.

1933

British physicist Arthur Eddington publishes *The Expanding Universe,* in which he lays out his theory that the universe is constantly increasing in size.

German-born U.S. astronomer Walter Baade suggests that supernovas develop into neutron stars after exploding.

1937

Austrian astronomer Marietta Blau examines cosmic radiation, using a photographic plate.

U.S. radio engineer Grote Reber builds the first radio telescope. It has a parabolic reflector 9.4 m/31 ft in diameter and begins service in Wheaton, Illinois.

1940

Belgian astronomer Marcel Gilles Jozef Minnaert publishes *Photometric Atlas of the Solar Spectrum,* a standard reference text providing measurements of the absorption lines from 3,332 angstroms to 8,771 angstroms.

1942

Radar operators in the British army detect, for the first time, radio emissions from the sun.

U.S. radio engineer Grote Reber makes the first radio maps of the sky, locating individual radio sources.

1944

Dutch astronomer Hendrik van de Hulst predicts that cosmic hydrogen will emit line radiation at 21 cm/8.3 in.

1945

Hungarian astronomer Zoltán Bay and the U.S. Army Signal Corps Laboratory at Fort Monmouth, New Jersey, receive radar echoes from the moon.

Hungarian scientist Lajos Jánossy investigates cosmic radiation.

1946

Cygnus A, the first radio galaxy (that is, a galaxy that is a strong source of electromagnetic waves of radio wavelengths) and the most powerful cosmic source of radio waves, is discovered.

British physicists Edward Appleton and Donald Hay discover that sunspots emit radio waves.

1948

June 3. The 5-m/200-in Hale reflector telescope is opened at Mount Palomar Observatory, near Pasadena, California; it remains the world's largest and most powerful telescope until 1974.

Austrian-born British mathematician Hermann Bondi, English astronomer and cosmologist Fred Hoyle, and Austrian-born U.S. astronomer Thomas Gold publish *The Steady State Theory of the Expanding Universe,* in which they argue that the universe is constantly expanding, but also maintaining a constant density through the continual creation of new stars and galaxies at a rate equal to the rate at which old ones become unobservable because of their increasing distance.

Dutch-born U.S. astronomer Gerard Kuiper discovers and photographs Miranda, the fifth moon of Uranus.

1949

German-born U.S. astronomer Walter Baade discovers the close-approach asteroid Icarus; except for comets, it has the most eccentric orbit of any body in the solar system and passes closest to the sun (28 million km/18 million mi).

1950

Cape Canaveral, Florida, is established as a rocket assembly and launching facility.

Dutch astronomer Jan Hendrik Oort proposes that comets originate in a vast cloud of bodies (the Oort cloud) that orbits the sun at a distance of about 1 light-year.

1951

September 20. The U.S. Air Force makes the first successful recovery of animals from a rocket flight when a monkey and eleven mice are recovered from a flight to an altitude of 72,000 m/236,000 ft.

Dutch-born U.S. astronomer Gerard Kuiper proposes the existence of a ring of small, icy bodies orbiting the sun beyond Pluto, thought to be the source of comets. They are discovered in the 1990s and named the Kuiper belt.

U.S. physicist Edward Purcell discovers line radiation (radiation emitted at only one specific wavelength) at 21 cm/8.3 in emitted by hydrogen in space, as predicted by Dutch astronomer Hendrick van de Hulst in 1944. The discovery allows the determination of the distribution of hydrogen clouds in galaxies and the speed of the Milky Way's rotation.

1952

U.S. astronomers Horace Babcock and Harold Babcock develop the solar magnetograph, a device for making detailed measurements of the sun's magnetic field.

1954

May 21. The Central Observatory of the Soviet Academy of Sciences, near Leningrad, USSR, opens.

1955

April. U.S. communications engineer John Pierce analyzes various types of satellites and shows the potential of using Earth's gravity to control the altitude and orientation of satellites in geosynchronous orbit. It leads to the launch of the U.S. communications satellite *Telstar 1*.

May 5. U.S. astronomer B. F. Burke announces the discovery that Jupiter emits radio waves.

September. The National Geographic Society, pointing to blue-green patches in photographs of the Martian surface, renews speculation that life exists on Mars.

Both the United States and the USSR announce that they will attempt launching satellites during the International Geophysical Year (1957–1958).

British radio astronomer Martin Ryle builds the first radio interferometer. Consisting of three antennae spaced 1.6 km/1 mi apart, it increases the resolution of radio telescopes, permitting the determination of the diameter of a radio source and the separation of two closely spaced sources.

French astronomer Audouin Dolfus ascends 7.2 km/4.5 mi above Earth in a balloon to make photoelectric observations of Mars.

1957

August 19. U.S. astronomers, using a 33-cm/12-in telescope on board the uncrewed balloon-telescope *STRATOSCOPE I,* take the first clear photographs of the sun from 24,384 m/80,000 ft.

October 4. The USSR launches the first artificial satellite, *Sputnik 1,* to study the cosmosphere. It weighs 84 kg/184 lb and circles Earth in 1.6 hours, inaugurating the space age.

November 3. The USSR places *Sputnik 2* in orbit carrying a dog, Laika. It is the first vehicle to carry a living organism into orbit. Laika dies in space.

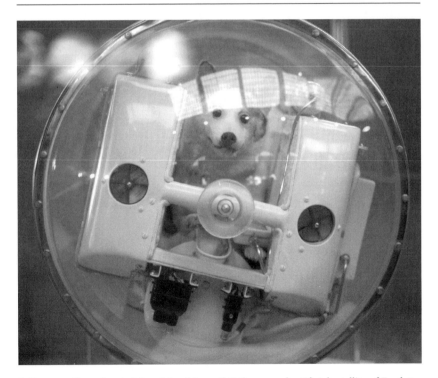

Laika peers from the airtight cabin of *Sputnik 2,* the second artificial satellite of Earth to be launched by the Soviets. Laika was the first living creature to orbit Earth. (Corbis/ Marc Garanger)

The Jodrell Bank radio telescope, located in Cheshire, England, and designed by British radio astronomer Bernard Lovell, begins operating. The first large radio telescope, it has a 76-m-/250-ft-diameter reflector, which can be rotated horizontally at 20° per minute and vertically at 24° per minute. Lovell uses it to track the Soviet satellite *Sputnik 1*, launched October 4.

The United States launches the first Jupiter-C three-stage rocket. It has a recoverable nose cone and is used for the first flights to carry live animals.

1958

January 31. The U.S. Army launches *Explorer 1* into Earth orbit. The first U.S. satellite, it is used to study cosmic rays.

May 15. The USSR places *Sputnik 3* in orbit. It contains the first multipurpose space laboratory and transmits data about cosmic rays, the composition of Earth's atmosphere, and ion concentrations.

May. Using data from the *Explorer* rockets, U.S. physicist James Van Allen discovers a belt of radiation around Earth. Now known as the Van Allen belts (additional belts were discovered later), they consist of charged particles from the sun trapped by Earth's magnetic field.

July 29. The United States creates the National Aeronautics and Space Administration (NASA) for the research and development of vehicles and activities involved in space exploration.

August 27. The USSR launches the satellite *Sputnik 5*, which carries two dogs to a height of 450 km/279 mi.

October 11. The United States launches the space probe *Pioneer 1* into orbit.

November 3. Soviet astronomer Nikolai Kozyrev observes the first volcanic eruption on the moon, a gaseous eruption in the crater Alphonsus.

November. The United States launches *Atlas* from Cape Canaveral, Florida. It is a single-stage rocket and has a range of 14,400 km/9,000 mi. It was originally designed as an intercontinental ballistic missile (ICBM).

December 18. *PROJECT SCORE* (Signal Communications by Orbiting Relay Equipment), the first U.S. communications satellite, is launched. It functions for thirteen days, relaying messages stored on magnetic tape.

Radio astronomers receive the reflection of radio waves from Venus. Similar echoes from other objects in the solar system allow an accurate measurement of their distance.

1959

January 2. The USSR launches *Lunik 1*. The first spacecraft to escape Earth's gravity, it passes within 6,400 km/4,000 mi of the moon.

February 28. The U.S. Air Force launches *Discoverer 1* into a low polar orbit, on a test flight. Later *Discoverer* missions photograph the entire surface of Earth every twenty-four hours, with the exposed film returned to Earth in its ejectable capsule.

1959

February. The U.S. Navy launches *Vanguard 2*, the first weather satellite.

March 3. The United States launches the moon probe *Pioneer 4*; it passes within 59,000 km/37,000 mi of the moon.

April. NASA selects a pool of nine military test pilots to compete to be the first U.S. astronaut to orbit Earth.

April. The United States launches *Discoverer 2*, a second military reconnaissance satellite.

May. The U.S. Army sends two monkeys some 500 km/300 mi into space. They are recovered, unharmed, in the Caribbean Sea.

August 7. NASA launches the U.S. space probe *Explorer 6*. It investigates the Van Allen radiation belt discovered in 1958 by *Explorer 1* and takes the first television pictures of Earth's cloud cover. The United States launches ten other satellites during the year.

September 14. The Soviet spacecraft *Luna 2* (launched September 12) becomes the first spacecraft to strike the moon.

October 7. The Soviet *Luna 3* (launched October 4) takes the first photographs of the far side of the moon.

British radio astronomer Martin Ryle and colleagues publish the *Third Cambridge Catalogue,* a catalog of radio sources that leads to the discovery of the first quasar.

The administration of U.S. President Dwight D. Eisenhower announces preliminary plans for a joint U.S.-USSR space venture.

U.S. astronomer Harold Babcock discovers that the sun periodically reverses its magnetic polarity.

1960

March 11. The United States launches *Pioneer 5* to study the solar winds between the planets.

April 1. The United States launches *TIROS 1* (Television and Infra-Red Observation Satellite). A weather satellite, it is equipped with television cameras, infrared detectors, and videotape recorders. Along with subsequently launched *TIROS* satellites, it provides a worldwide weather-observation system.

April. The U.S. Thor-Able-Star corporation launches *Transit 1-B,* a satellite expected to facilitate navigation.

May 15. The USSR launches *Spacecraft I.* Weighing 4,540 kg/10,000 lb, it is the first vehicle large enough to contain a human passenger.

May. The United States launches *Midas 2,* a satellite designed to provide early warning of a missile attack.

August 12. NASA launches *Echo 1,* a 30-m/100-ft aluminum-coated balloon used as a passive communications satellite to reflect radio waves. It remains in orbit for eight years and is a conspicuous object in the night sky.

August 18. The United States launches the satellite *Discoverer 14.* A U.S. Air Force C-119 transportation plane recovers its capsule in midair over the Pacific.

August 21. The USSR safely retrieves *Spacecraft II,* which has two dog passengers.

British radio astronomer Martin Ryle develops the synthesized-aperture interferometer. Using two or more movable antennae mounted on rails, it greatly improves the resolution of radio telescopes, allowing the mapping of such distant radio sources as quasars.

December. There are now twenty artificial satellites in orbit.

1961

January. The United States sends a second primate, this time a chimpanzee, 249 km/155 mi into space aboard a Project Mercury spacecraft. The chimp is successfully recovered.

April 12. Soviet cosmonaut Yuri Gagarin, in *Vostok 1,* is the first person to enter space. His flight lasts 1.8 hours.

May 5. U.S. astronaut Alan Shepard in the Mercury capsule *Freedom 7* makes a 14.8-minute suborbital flight. He is the first U.S. astronaut into space.

May 21. U.S. President John F. Kennedy commits the country to landing someone on the moon and returning them safely to Earth before the decade is out.

May. U.S. President John F. Kennedy asks Congress to allocate close to $2 billion for space exploration as part of his pledge of May 21 to land a man on the moon before the end of the decade.

August 7. Soviet cosmonaut Gherman Titov, the second cosmonaut to be launched into space, completes seventeen orbits in 25.5 hours in *Vostok 2* and becomes the first person to spend more than a day in space.

October 27. The United States launches the first two-stage *Saturn 1* rocket. The first rocket specifically designed for spaceflight, it is used to launch the *Apollo* spacecraft.

The Soviet space probe *Venera 1* passes within 99,000 km/62,000 mi of Venus but fails to transmit data due to a telemetry failure.

U.S. astronaut Alan Shepard in the Mercury capsule *Freedom 7*.

1962

January 26. NASA launches *Ranger 3,* a spacecraft designed to photograph and then explore the lunar surface. But *Ranger 3* misses its rendezvous with the moon and ends up in a solar orbit.

February 20. U.S. astronaut John Glenn, in the Project Mercury capsule *Friendship 7,* becomes the first U.S. astronaut to orbit Earth. He makes three orbits.

March 7. The United States launches the *Orbiting Solar Observatory* (OSO). The first of a series of solar observatories, it collects and transmits data on the sun's electromagnetic radiation.

April. Like its predecessor, *Ranger 3,* NASA's lunar probe *Ranger 4* malfunctions and crashes on the moon.

April 23. *Ranger 4* becomes the first U.S. spacecraft to hit the moon.

April 26. The United States and Britain launch Earth satellite *Ariel.* Designed to study the ionosphere, it is the first international cooperative launch.

June 14. The European Space Research Organization is established in Paris, France.

July 10. NASA launches the communications satellite *Telstar* for the American Telephone and Telegraph Company (AT&T) from Cape Canaveral, Florida. Weighing 77 kg/170 lb and orbiting Earth every 2.63 hours, it is designed to receive a signal from the ground, amplify it, and then relay it to another ground station. Live television pictures of the chairman of AT&T are transmitted from Andover, Maine, to Goonhilly Down, Cornwall, England, and Brittany, France. Transmissions last only fifteen minutes per orbit, but they are the first to connect the television networks of Europe and North America.

August 26. The United States launches the space probe *Mariner 2.* It makes a flyby of Venus (December 14), passing within 34,000 km/21,000 mi of the planet's surface, and takes measurements of temperature and atmospheric density.

November 1. The USSR launches the *Mars 1* probe to Mars; it flies in the right direction but transmits no data because of a radio failure.

December 5. The United States and the USSR sign an agreement on co-operation for the peaceful use of outer space.

December 13. NASA launches the experimental telecommunications satellite *Relay 1* for the Radio Corporation of America (RCA).

Italian astronomers Riccardo Giacconi, Herbert Gursky, Frank Paolini, and Bruno Rossi discover the first astronomical X-ray source—in Scorpio.

1963

May 15–16. U.S. astronaut Gordon Cooper, in *Faith 7,* the last of the Project Mercury missions, completes a twenty-two-orbit mission in 34.3 hours (a record) and then makes a manual landing when his automatic controls fail.

May 20. The United States launches *MIDAS,* a satellite containing a 20-kg/44-lb belt of copper wires into a 3,000-km/1,865-mi polar orbit. The wires are to serve as a reflective umbrella with which to test-relay radio and microwave signals from coast to coast.

June 16. Soviet cosmonaut Valentina Tereshkova, the first woman in space, is launched into a three-day Earth-orbit flight aboard *Vostok 6* to study the problem of weightlessness.

October 26. Soviet leader Nikita Khrushchev states that the USSR will not race the United States to place a person on the moon.

An international team of astronomers discovers the first quasar (3C 273), an extraordinarily distant object that is brighter than the largest-known galaxy yet has a starlike image.

The Arecibo radio telescope in Puerto Rico begins operation; its 305-m/1,000-ft reflector is built into a naturally occurring parabola and is the largest single-reflector telescope in the world.

The European Launcher Development Organization is formed to develop space-launch vehicles. The test launch site is in Woomera, Australia.

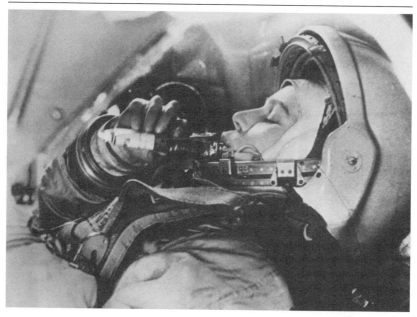

Valentina Tereshkova practices eating in flight simulations for launch into space (ca. 1963). (Corbis-Hulton-Deutsch Collection)

1964

July 28. The United States launches the spacecraft *Ranger 7* from Cape Kennedy (the name by which Cape Canaveral was known from 1963 to 1973), Florida; it succeeds in obtaining close-up photographs of the moon's surface before crashing on July 31.

August 28. The United States launches *Nimbus 1,* the first meteorological satellite, into a polar orbit. It replaces the *TIROS* satellites.

October 12–13. The USSR launches *Voskhod 1,* the first spacecraft to have a crew of three.

November 28. The United States launches *Mariner 4* to Mars. It will pass within 9,800 km/6,100 mi of the planet on July 14, 1965, and relay the first close-up photographs of the surface, as well as information on the Martian atmosphere.

English astronomer and cosmologist Fred Hoyle and Indian astronomer Jayant Narlikar propound a new theory of gravitation that solves the problem of inertia.

1965

March 18. Soviet cosmonaut Alexsi Leonov leaves the spacecraft *Voskhod 2* and floats in space for twenty minutes—the first space walk.

March 23. U.S. astronauts Virgil I. ("Gus") Grissom and John Young are launched aboard *Gemini 3*. It is the first U.S. space mission with a crew of two.

June 3–7. U.S. astronaut Edward White, during the *Gemini 4* space mission, demonstrates the ability of humans to function in outer space when he makes a twenty-two-minute space walk, the first by a U.S. astronaut. He is also the first to use a personal-propulsion pack during the walk.

July. The Soviet spacecraft *Zond 3* relays close-up photographs of 7.8 million km²/3 million mi² of the moon's surface.

August 21–29. The U.S. spacecraft *Gemini 5* completes 120 orbits in eight days. It is the longest spaceflight taken to date and demonstrates the ability of humans to adapt to weightlessness.

November 26. The French launch their first satellite, *A-1 Asterix*.

December 12. The U.S. spacecraft *Gemini 6* (launched December 4) comes to within 0.3 m/1 ft of *Gemini 7* (launched December 5). They are the first spacecraft to rendezvous with each other.

December 16. The United States launches *Pioneer 6* into solar orbit. It will relay information about the solar wind, cosmic rays, and the tail of Comet Kohoutek.

German-born U.S. radio engineer Arno Penzias and U.S. radio astronomer Robert Wilson detect microwave background radiation in the universe and suggest that it is the residual radiation from the Big Bang.

1966

February 3. The Soviet spacecraft *Luna 9* (launched January 31) makes the first soft landing on the moon and transmits photographs and soil data for three days.

February. U.S. astronauts Elliot See Jr. and Charles Basset II perish when their T-38 aircraft crashes near St. Louis, Missouri, en route to a training session.

March 1. The Soviet probe *Venera 3* (launched November 16, 1965) crash-lands on Venus, the first artificial object to land on another planet.

March 16. U.S. astronauts Neil Armstrong and David Scott, aboard *Gemini 8*, achieve the first linkup of a crewed spacecraft with another object, an *Agena* rocket.

June 2. NASA spacecraft *Surveyor I* (launched May 30) makes the first U.S. soft landing on the moon and transmits more than ten thousand photographs of the lunar surface.

August 10. The U.S. spacecraft *Lunar Orbiter 1*, launched this day, enters moon orbit and transmits pictures of the dark side. It is the first of five uncrewed spacecraft that photograph the moon to select sites for the *Apollo* missions and to make detailed lunar maps.

November 11. The United States launches *Gemini 12*, the last of the *Gemini* two-person space missions. It makes the first fully automatically controlled reentry.

Astronomers at the U.S. Naval Research Laboratory in Washington, D.C., discover powerful X rays emitted from within the constellation Cygnus.

1967

January 27. A treaty banning nuclear weapons from outer space is signed by sixty countries, including the United States and the USSR. It will be effective from October 10.

January 27. Three U.S. astronauts, Virgil I. ("Gus") Grissom, Edward White, and Roger B. Chaffee, die in a fire during a countdown rehearsal on the *Apollo 1* spacecraft at Cape Kennedy, Florida. They are the first human casualties of the U.S. space program.

April 17. The U.S. spacecraft *Surveyor 3* is launched and soft lands on the moon, where it conducts sampling experiments on the lunar soil. It is subsequently visited by astronauts from the *Apollo 12* mission.

April 24. Soviet cosmonaut Vladimir Komarov dies during the descent of his *Soyuz 1* spacecraft when his parachute fails to open properly. He is the first fatality during a spaceflight and the first of the Soviet space program.

July 7. British astronomers Jocelyn Bell and Antony Hewish at the Mullard Radio Astronomy Observatory, Cambridge, discover the first pulsar (announced in 1968). A new class of stars, they are later shown to be collapsed neutron stars emitting bursts of radio energy. The Crab Nebula supernova remnant is discovered to be a pulsar the following year.

September 8. The United States launches the spacecraft *Surveyor 5*. It measures the proportions of chemicals in the lunar soil.

October 18. The Soviet spacecraft *Venera 4* (launched June 12) lands on Venus, the first soft landing on another planet; its instrument-laden capsule transmits information about Venus's atmosphere.

October 19. The U.S. spacecraft *Mariner 5* (launched June 14) passes within 4,000 km/2,500 mi of the surface of Venus, transmitting data on the planet's atmosphere and magnetic field as it does so.

October 22–29. The Soviet spacecraft *Cosmos 186* and *Cosmos 188* complete the first automatic docking in space.

November 7. The U.S. spacecraft *Surveyor 6* photographs one area of the moon, then lifts off, repositions itself 2.4 m/8 ft away, and resumes photographing. It is the first liftoff from an extraterrestrial body.

November 9. The United States launches first three-stage *Saturn V* rocket, carrying *Apollo 4*. Weighing more than 2,700 metric tons/3,000 tons, it is used to lift the *Apollo* missions to the moon.

A tank containing 100,000 gallons of cleaning fluid is installed in a former gold mine in South Dakota to detect neutrinos (uncharged elementary particles) from the sun.

The U.N. Outer Space Treaty that bans military activities in space is ratified by sixty-three member nations of the United Nations.

1968

March 7. The United States launches the *Explorer 36* satellite, which tests the feasibility of using lasers to communicate in space.

June 12. The first radar observations of an asteroid are made when the asteroid Icarus approaches within 6 million km/3.7 million mi of Earth.

September 14–21. The Soviet spacecraft *Zond 5* flies around the moon and returns to Earth—the first spacecraft to do so.

October 11–22. Launched October 11, *Apollo 7*, the first U.S. *Apollo* space mission with a crew, tests the command module that will be used on subsequent flights to the moon, during 163 orbits of Earth. The crew delivers the first live television program from space on October 13.

December 21–27. Launched December 21, the U.S. *Apollo 8* spacecraft flies around the moon.

1969

January 16. Two cosmonauts aboard the Soviet spacecraft *Soyuz 5* (launched January 15) dock and transfer to *Soyuz 4* (launched January 14). Locked together for four hours, they form the first experimental space station.

February–March. The U.S. space probe *Mariner 6* (launched February 24) passes within 3,410 km/2,131 mi of the surface of Mars. *Mariner 7* (launched March 27) photographs the Martian landscape and makes thermal maps of the planet and analyzes its atmosphere.

March 3–13. U.S. astronauts on the *Apollo 9* mission (launched March 3) test the lunar module in Earth orbit.

May 22. Two astronauts aboard the U.S. *Apollo 10* spacecraft (launched May 18) transfer into the lunar module and descend to within 14.5 km/9 mi of the moon's surface.

July 16–24. The U.S. moon-shot mission *Apollo 11*, launched July 16, takes place. On July 20, U.S. astronaut Neil Armstrong becomes the first person

Buzz Aldrin walks on the moon (20 July 1969). (NASA)

to walk on the moon, famously saying: "That's one small step for man, one giant leap for mankind." He and Buzz Aldrin also install and operate the first moon seismograph at Tranquility base, spending a total of 21.6 hours on the moon's surface, while Michael Collins remains orbiting the moon in the command module.

November 22. The lunar module from the U.S. spacecraft *Apollo 12* (launched November 14) lands on the moon in the Ocean of Storms. The crew collects 34 kg/75 lb of moon rocks, inspects *Surveyor 3*, which had landed nearby more than two years earlier, and uses a radioisotope-fueled generator to power experiments.

A multichannel spectrometer is installed at the Mount Palomar Observatory, near Pasadena, California. It permits the rapid and accurate collection of data through the simultaneous observation of thirty-two wavelength bands.

The Sacramento Peak Observatory, Sunspot, New Mexico, becomes operational. All of the air from its solar telescope, which is 76 cm/30 in in diameter and 54.9 m/180 ft long, has been evacuated to prevent it from overheating.

The Soviet spacecraft *Soyuz 6*, *Soyuz 7*, and *Soyuz 8* are launched on successive days (October 11, 12, and 13, respectively). The crews conduct welding experiments in space.

1970

April 13–17. NASA narrowly averts a disaster aboard the moon-bound spaceship *Apollo 13* (launched April 11) after a canister of liquid oxygen explodes in the command module. The crew of James Lovell, John Swigert, and Fred Haise enter the lunar module, which they use as a "lifeboat" to return safely to Earth.

June. The Soviet spacecraft *Soyuz 9*, launched June 1, remains in orbit for eighteen days, establishing an endurance record.

September 12–21. The Soviet uncrewed spacecraft *Luna 16*, launched September 12, lands on the moon, collects soil samples in a sealed container, and returns to Earth.

November. The Soviet spacecraft *Luna 17* (launched November 10) lands on the moon and deploys the remotely controlled roving moon vehicle *Lunokhod 1*.

NASA's *Orbiting Astronomical Observatory* (launched December 7, 1968) and the *Orbiting Geophysical Observatory* (launched March 3, 1968) detect hydrogen in the tail of a comet.

The United States launches the *Small Astronomy Satellite* (SAS). It catalogs X-ray sources and leads to the development of the *High-Energy Astronomy Observatory* (HEAO).

The Soviet spacecraft *Venera 7* transmits information from the surface of Venus.

1971

The Effelsberg radio telescope near Bonn, Germany, begins operating; its 100-m/328-ft movable dish is the largest fully steerable dish in the world.

January. The international *Intelsat 4* is launched; it can handle three thousand to nine thousand telephone circuits or twelve color-television channels simultaneously.

January 31–February 8. The U.S. spacecraft *Apollo 14,* launched January 31, under the command of astronaut Alan Shepard, reaches the moon on February 5 and returns to Earth three days later, having collected 43 kg/ 95 lb of lunar rock.

April 19. The USSR launches the 15-m-/50-ft-long *Salyut 1* space station. It is visited by a three-person crew June 7–29, but the cosmonauts die during their return to Earth when a faulty valve causes their capsule to lose pressure. The station reenters Earth's atmosphere six months later.

May 8. The uncrewed U.S. spacecraft *Mariner 8* plummets into the Atlantic Ocean after the failure of a booster rocket.

July 26–August 7. On a lunar mission aboard *Apollo 15* (launched July 26), U.S. astronauts David R. Scott and James R. Irwin use the spacecraft's Lunar Rover vehicle to explore 27 km/17 mi of the moon's surface.

November 24. The U.S. space probe *Mariner 9* (launched May 30) becomes the first artificial object to orbit another planet (Mars); it transmits 7,329 photographs of the planet and its two moons, Deimos and Phobos.

November 27. The Soviet *Mars 2* lander (launched in May 19) crashes on Mars, returning no data.

December 2. A lander from *Mars 3* (launched May 28) successfully lands on Mars but its transmitter goes dead after twenty seconds.

British theoretical physicist Stephen Hawking suggests that, after the Big Bang, "mini" black holes—each no bigger than a proton but containing a billion metric tons of mass—were formed and that they are governed by both the laws of relativity and of quantum mechanics.

December. Scientists designate Cignus X-1 as the first probable black hole.

1972

January 5. U.S. President Richard Nixon authorizes a $5.5 billion six-year program to develop plans for a spaceship capable of undertaking multiple missions, thereby launching the space-shuttle program.

March 3. The uncrewed U.S. spacecraft *Pioneer 10* takes off toward Jupiter.

April 16–27. On the U.S. *Apollo 16* lunar mission (launched April 16), astronauts Charles Duke and John Young gather 97 kg/214 lb of lunar soil and rock during a record seventy-one hours and two minutes on the moon.

July 22. The Soviet craft *Venera 8* (launched March 27) makes a soft landing on Venus and transmits valuable data from the surface.

July 23. The U.S. launches *Landsat 1,* the first of a series of satellites for surveying Earth's resources from space.

December 7–19. Astronaut Eugene Cernan commands *Apollo 17* (launched December 7) on the last U.S. crewed mission to the moon.

1973

April 6. The uncrewed U.S. spacecraft *Pioneer 11* rockets toward Jupiter.

May 14. The United States launches the first *Skylab* space station. It contains a workshop for carrying out experiments in weightlessness, an observatory for monitoring the sun, and cameras for photographing Earth's surface. *Skylab* is subsequently visited by three three-person U.S. crews, and astronauts make observations of the sun, manufacture superconductors, and conduct other scientific and medical experiments. The third mission (November 16, 1973–February 8, 1974) lasts a record eighty-four days and gathers data about long spaceflights.

May 25. The U.S. *Skylab 2* transports a team of astronauts on a repair mission to the original U.S. *Skylab* space station, damaged during its launch on May 14.

July 28. *Skylab 3* conveys a team of U.S. astronauts to the original *Skylab* space station to make further repairs and conduct experiments.

November 3. The United States launches the uncrewed spacecraft *Mariner 10*, bound for the planet Mercury via Venus.

December. The U.S. probe *Pioneer 10* (launched March 3, 1972) passes within 130,000 km/81,000 mi of Jupiter, taking hundreds of photographs. It is destined to travel beyond the solar system, leaving it on June 13, 1983.

Comet Kohoutek is first observed by Czech astronomer Lubos Kohoutek.

The Nicholas Mayall Telescope begins operation at Kitt Peak National Observatory, near Tucson, Arizona; its 4-m/158-in dish makes it the second-largest optical telescope in the world.

1974

May 17. The United States launches the world's first *Synchronous Meteorological Satellite* (SMS). There had been many weather satellites before, but this was the first in geosynchronous orbit, that is, facing the same part of the Earth continuously.

September. U.S. astronomer Charles T. Kowal announces the discovery and naming of Leda, the thirteenth moon of Jupiter.

December. The U.S. probe *Pioneer 11* (launched in April 1973) flies to within 42,000 km/26,000 mi of Jupiter and photographs its polar regions.

British theoretical physicist Stephen Hawking suggests that black holes emit subatomic particles until their energy is diminished to a point at which they explode.

British radio astronomers Martin Ryle and Antony Hewish receive the Nobel Prize for Physics for their work in radio astronomy.

The Global Atmospheric Research Program (GARP) is announced. An international project organized by the World Meteorological Organization and the International Council of Scientific Unions, its aim is to provide a greater understanding of the mechanisms of the world's weather

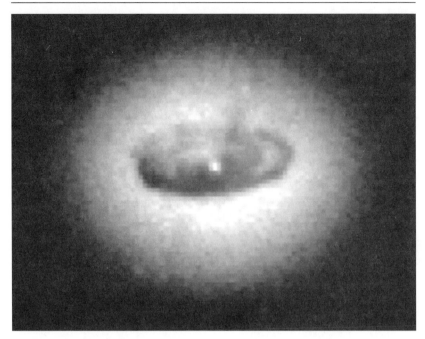

Digital image of a galaxy core. (NASA)

by using satellites and by developing a mathematical model of Earth's atmosphere.

October 15. The British X-ray satellite *Ariel 5* is launched by NASA from the San Marco launch platform in the Indian Ocean. Ariel 5 monitored the X-ray sky. It discovered "long period" pulsars, i.e., pulsars with pulse periods of the order of minutes.

The U.S. probe *Mariner 10* (launched November 3, 1973) photographs the upper atmosphere of Venus (February) and then takes the first photographs of the surface of Mercury (March and September), flying within 740 km/460 mi of the planet's surface.

1974–1975

France and West Germany launch two *Symphonie* satellites, *Symphonie A* on December 12, 1974, and *Symphonie B* on August 27, 1975. Placed in geosynchronous orbit over West Africa, they provide Africa with telecommunications services to western Europe and the Americas.

1975

March 15. The U.S.-German space probe *Helios 1* (launched December 10, 1974) passes the sun at a distance of 45 million km/28 million mi and returns information about the sun's magnetic field and solar wind.

July 15. The launch of the Soviet spacecraft *Soyuz 19* signals the start of a joint U.S.-Soviet space mission. U.S. and Soviet astronauts meet in space on July 17 when *Soyuz 19* docks with its U.S. counterpart, *Apollo 18.*

August 1. The European Space Agency is founded in Paris, France, to undertake research and develop technologies for use in space.

October 22–25. The Soviet spacecraft *Venera 9* and *10,* launched June 8 and June 14, respectively, land on Venus and transmit the first pictures from the surface of another planet.

October 16. The United States launches the first *Geostationary Operational Environmental Satellite* (GOES), which provides twenty-four-hour coverage of U.S. weather.

January 22. The U.S. launches *Landsat 2.* It is positioned 180° from *Landsat 1* (launched July 23, 1972); together, the two offer regular images of Earth, including the capability to provide a view of the same geographic area with the same sun angle every nine days.

French researcher Antoine Labeyrie develops an interferometer (a device using the interference of waves to make precise measurements) that can distinguish between two objects 1 m/3.3 ft apart at a distance of 60,000 km/37,284 mi; it leads to even more powerful interferometers that prove useful in making precise measurements of celestial bodies.

The Anglo-Australian Telescope begins operation at Siding Spring Mountain, New South Wales, Australia; its 3.9-m/154-in reflector makes it the third-largest optical telescope in the world and the largest in the Southern Hemisphere.

U.S. radio astronomers Russell Hulse and Joseph Taylor identify PSR1913+16 as a binary star (a pair of stars in orbit around each other); it loses energy at a rate that Einstein's theory of general relativity predicts for the emission of gravitational waves—ripples in the structure of spacetime, which may occur singly or as continuous radiation.

The USSR and the United States simultaneously develop beam weapons that produce beams of protons that destroy solid targets such as satellites.

1976

January 15. The U.S.-German space probe *Helios 2* is launched; able to withstand extremely high temperatures, it passes within 43.4 million km/ 27 million mi of the sun and relays data about cosmic rays, the sun's magnetic field, and the solar wind.

May 4. The U.S. launches *Lageos* (Laser Geodynamic Satellite); it uses laser beams to make precise measurements of Earth's movements in an attempt to improve the prediction of earthquakes. Placed in an orbit 9,321 km/5,793 mi high, it is expected to remain aloft for eight million years.

Astronomers at Harvard College Observatory in Cambridge, Massachusetts, discover bursts of X rays coming from a star cluster 30,000 light-years from Earth.

The 6-m/236-in UTR-Z telescope is completed at Zelenchukskaya, USSR (now Russia); it is the largest reflecting telescope in the world.

July 20, September 3. The U.S. spacecraft *Viking 1* and *Viking 2* (launched in 1975) soft land separately on Mars. They make meteorological readings of the Martian atmosphere and search for traces of bacterial life which prove inconclusive.

NASA launches three *Marisat* (Maritime Satellite Systems) satellites; they improve ship-to-shore communications.

1977

March. U.S. astronomer James Elliot and several groups of other U.S. astronomers discover rings around Uranus when the planet occludes a relatively bright star.

August 12. The National Aeronautics and Space Administration launches the *High Energy Astronomy Observatory 1* (HEAO 1). It generates a catalog of X-ray sources.

U.S. astronomer Charles Kowal discovers the interplanetary body Chiron; at least 200 km/120 mi in diameter, 182 km/113 mi in diameter, it is initially thought to be an asteroid but is later identified as a giant planetary nucleus.

U.S. futurist Peter E. Glaser proposes that energy be beamed to Earth by a solar collector placed in geostationary orbit; with an area of 100 km^2/40 mi^2, it could supply about 10 gigawatts (10^9 watts) of electricity.

1978

The Nobel Prize for Physics goes to German-born U.S. radio engineer Arno Penzias and U.S. radio astronomer Robert Wilson for their discovery of microwave background radiation, which supports the Big Bang theory. They share the prize with Soviet physicist Pyotr Kapitza, who is awarded for his invention and application of the helium liquefier.

January 26. The U.S.-European *International Ultraviolet Explorer* is launched into geostationary orbit; it provides data on ultraviolet sources in outer space.

Satellite data begin to be useful in the discovery of new oil deposits.

March 5. The U.S. launches *Landsat 3*, completing the initial series of land-surveying satellites.

June 15–Novemberr 2. Two Soviet cosmonauts spend a record 139.58 days in space.

June 22. U.S. astronomer James W. Christy discovers Charon, a moon orbiting Pluto.

June 27. The United States launches the *Seasat 1* satellite to measure the temperature of sea surfaces, wind and wave movements, ocean currents, and icebergs; it operates for ninety-nine days before its power fails.

December. The Soviet spacecraft *Venera 11* and *12* (launched September 9, 1977, and September 14, 1977) soft land on Venus (December 21 and December 25, respectively) and collect data on the lower atmosphere.

December. The U.S. space probes *Pioneer-Venus 1* (launched May 20) and *2* (launched August 8) go into orbit around Venus, the first relaying information about the atmosphere, the second taking radar photographs of the surface that reveal huge mountains and basins.

Miniature nuclear reactors begin to be made to power radar satellites in the USSR.

1979

February 25–August 19. Soviet cosmonauts Vladimir Lyakhov and Valeri Ryumin set a new record for time spent in space—175.02 days.

July. The U.S. space station *Skylab 1* falls back to Earth after traveling 140 million km/87 million mi in orbit since 1973.

December 4. The European Space Agency launches its first *Ariane* rocket from the Guiana Space Center in Kourou, French Guiana; it is designed to deploy satellites into orbit.

The 3-m/120-in U.S. Infrared Telescope Facility (IRTF) telescope, the 3.8-m/150-in British Infrared Telescope (UKIRT), and the 3.6-m/142-in Canada-France-Hawaii Telescope (CFHT) begin operation on Mauna Kea, Hawaii.

The IRAM Array telescope begins operation at Plateau de Bruce, France; its four 15-m/19.2-ft dishes make it the largest millimeter telescope in the world.

The Multiple Mirror Telescope begins operation on Mount Hopkins, Arizona; it focuses the light from six 180-cm/70-in telescopes to form one image, giving the light-gathering power of a single 4.5-m/15.7-ft telescope; it becomes the prototype for larger optical telescopes.

The U.S. satellite *HEAO 3 (High Energy Astronomy Observatory*; later renamed the *Einstein Observatory)*, launched September 20, 1979, discovers numerous X-ray sources.

The Soviet space station *Salyut 6* (launched September 29, 1977) is the first to be docked by two other spacecraft.

September 1. The U.S. space probe *Pioneer 11* (launched in April 6, 1973) travels through the rings of Saturn to within 20,900 km/13,000 mi of the planet. The rings are found to be made of ice-covered rocks. Two additional rings and high-energy particles within Saturn's magnetosphere are also discovered.

U.S. astronomers R. Carswell and R. Weymann and British astronomer D. Walsh discover the first gravitational lens—a massive foreground object that bends the light from background objects.

U.S. astronomers John Eddy and Aram Boornazian announce that the sun is shrinking at a rate of 1.5 m/5 ft per hour.

U.S. space probes *Voyager 1* (launched September 5, 1977) and *2* (launched earlier, on August 20, 1977) fly past Jupiter on March 5, 1979, and July 9, 1979, respectively. *Voyager 1* discovers a ring around Jupiter and three moons.

1980

October 11. The Soviet cosmonauts Valeri Ryumin and Leonid Popov return to Earth after a record 185 days in space aboard *Salyut 6*.

November. 12. The U.S. space probe *Voyager 1* flies past Saturn within 124,000 km/77,000 mi; it discovers the planet's thirteenth, fourteenth, and fifteenth moons and transmits information about the planet, its moons, and its rings.

A thin layer of iridium-rich clay, about sixty-five million years old, is found around the world. U.S. physicist Luis Alvarez suggests that it was caused by the impact of a large asteroid or comet that threw enough dust into the sky to obscure the sun and cause the extinction of the dinosaurs.

U.S. astrophysicist Alan Guth proposes the theory of the inflationary universe—that the universe expanded very rapidly for a short time after the Big Bang.

The Multi-Element Radio-Linked Interferometer Network (MERLIN) radio telescope begins operation. Centered at Jodrell Bank, Cheshire, England, it eventually has five dishes measuring 25 m/82 ft, one measuring 32 m/105 ft, and one measuring 76 m/250 ft (the Lovell Telescope), making it the largest radio telescope in the world.

1981

The Very Large Array (VLA) radio telescope at Socorro, New Mexico, enters service; its twenty-seven 25-m/82-ft-diameter dishes are steerable and movable on railroad tracks and are equivalent to one dish 27 km/17 mi in diameter; together, they provide high-resolution radio images.

U.S. astronomer Uwe Fink and associates report the discovery of a thin atmosphere on Pluto.

April 12–14. The U.S. reusable space shuttle, using the orbiter *Columbia*, makes its first flight (second shuttle flight, November 12–14). Its landing is the first of a U.S. spacecraft on land.

August. The U.S. probe *Voyager 2* (launched August 20, 1977) records data on the atmospheres of Saturn and its moon Titan.

The most massive known star in the universe, R136, is discovered; it is twenty-five hundred times more massive than our sun and ten times as bright.

Space Shuttle *Columbia*, 12 April 1981. (NASA)

November 11. The orbiter *Columbia* makes the first deployment of a satellite from the U.S. space shuttle.

1982

Astronomers at Villanova University, Villanova, Pennsylvania, announce the discovery of rings around Neptune.

Astrophysicists at Groningen University in the Netherlands postulate the existence of a black hole at the center of the Milky Way.

September 9. The U.S. rocket *Conestoga 1* makes a suborbital flight; it is the first privately operated launch.

1983

June 25. The U.S.-British-Dutch *Infrared Astronomy Satellite* (IRAS) an orbiting observatory, is launched; it is designed to detect infrared radiation from objects in space and surveyed almost the entire infrared sky. It also finds the first evidence of planetary material around the star Vega outside our solar system.

June 13. The U.S. space probe *Pioneer 10* (launched March 3, 1977) becomes the first artificial object to leave the solar system.

June 18. Astronauts on board the U.S. space shuttle *Challenger* first use the Remote Manipulating Structure ("arm") to deploy and retrieve a satellite.

June 18–24. The *Challenger* mission (launched June 18) includes Sally Ride, the first U.S. woman to go into space.

September 30. Guion Bluford becomes the first African American to go into space, aboard the U.S. space shuttle.

Studies from the U.S. *Lageos* satellite (launched May 4, 1976) monitors slight movements in Earth's crust, which are used to help predict earthquakes.

The Nobel Prize for Physics is awarded jointly to Indian-born U.S. astrophysicist Subrahmanyan Chandrasekhar and U.S. astrophysicist

William Fowler for their studies of the life cycle of stars and the importance of nuclear reactions for the formation of chemical elements in the universe.

The Soviet space probes *Venera 15* and *16* (launched on June 2 and June 7) enter orbit around Venus on October 10 and October 14, respectively. *Venera 15* maps the surface of Venus.

1984

February 7. Two U.S. space-shuttle astronauts on board the *Challenger* make untethered space walks, using jet-propelled backpacks to move in space.

July 18. Soviet cosmonaut Svetlana Savitskaya becomes the first woman to walk in space.

November 12. U.S. astronauts on board the *Discovery* shuttle (launched November 8, 1984) use a Manned Maneuvering Unit (MMU) to retrieve two communications satellites.

U.S. astronomers at Cornell University, Ithaca, New York, report the discovery of eight infrared galaxies—thought to resemble primeval galaxies—located by the *Infrared Astronomy Satellite* (IRAS).

U.S. astronomers working in Chile photograph a partial ring system around Neptune.

1985

European, Japanese, and Soviet probes are launched to rendezvous with Halley's comet in 1986.

August 27–September 3. The U.S. space shuttle *Discovery* (launched August 27, 1985) deploys three satellites; the crew also retrieves, repairs, and redeploys an orbiting satellite.

November 27–December 4. The crew of the U.S. space shuttle *Atlantis* (launched November 27, 1985) undertakes construction exercises to develop skills for building a large orbiting space station.

1986

January 28. The U.S. space shuttle *Challenger* explodes shortly after take-off, killing the crew of seven and setting the U.S. space program back years.

January. The U.S. space probe *Voyager 2* (launched in 1977) passes within 81,000 km/50,600 mi of Uranus; photographs taken by the probe reveal ten unknown satellites and two new rings.

February 19. The USSR launches the core unit of the *Mir 1* space station; it is intended to be permanently occupied.

March 7–17. The thirtieth recorded appearance of Halley's comet. During these ten days, spacecraft from the USSR, Europe, and Japan fly by the comet. It comes closest to Earth, but is less bright, on February 5.

1987

February 8–December 29. Soviet cosmonaut Yuri Romanenko spends a record 326 days in the *Mir* space station.

February 23. Astronomers around the world observe a spectacular supernova in the Large Magellanic Cloud, the galaxy closest to ours, when a star (SN1987A) suddenly becomes a thousand times brighter than our own sun. It is the first supernova visible to the naked eye since 1604.

Harvey Butcher of Groningen University in the Netherlands estimates that the universe is younger than ten billion years.

Objects the size of planets are found orbiting the stars Gamma Cephei and Epsilon Eridani.

Radio waves are observed from 3C326—believed to be a galaxy in the process of formation.

The Britain-Netherlands James Clerk Maxwell Telescope (JCMT), operated by the Royal Observatory, based in Edinburgh, Scotland, begins operation on Mauna Kea, Hawaii; its 15-m/50-ft dish makes it the largest submillimeter telescope in the world.

The USSR launches the *Cosmos* satellite; it is the two-thousandth Soviet satellite.

The USSR launches the *Energiya* superbooster; it is the world's most powerful space launcher, with a thrust of 3 million kg/6.6 million lb.

1988

July. The USSR launches two *Phobos* space probes (July 7 and July 12) to study Phobos, one of the moons of Mars; *Phobos 1* is accidentally sent a "suicide" instruction and self-destructs.

September 19. Israel launches *Horizon,* its first satellite; it is used for geophysical studies.

September 29–October 3. The U.S. space shuttle *Discovery* (launched September 29) makes the first shuttle mission since the *Challenger* disaster more than two years previously.

November 15. The Soviet uncrewed space shuttle *Buran* (Blizzard) makes its inaugural flight, under radio control.

Simon Lilly of the University of Hawaii reports the location of a galaxy 12 billion light-years from Earth, adding to evidence about the date of the universe's formation.

1989

March 24. The United States launches the *Delta Star* "Star Wars" satellite; it successfully detects and tracks test missiles shortly after they are launched.

May 4. The U.S. space shuttle *Atlantis* launches the probe *Magellan* to map the surface of Venus using radar.

August 25. The U.S. space probe *Voyager 2* (launched August 20, 1977) reaches Neptune and transmits pictures; it discovers a great dark spot on the planet and six new moons.

October 18. The U.S. space shuttle *Atlantis* launches the probe *Galileo* to explore Jupiter. It will reach its destination in December 1995.

U.S. astronomers at Harvard College Observatory in Cambridge, Massachusetts, discover a large, thin sheet of galaxies, which they name the Great Wall; no current astronomical theory can explain its distinctive form.

Astronomers discover a river of gas at the center of the Milky Way, providing further evidence that a black hole, four million times as massive as the sun, exists at the center of our galaxy.

Star HD 114762 is discovered to have a planetlike body circling it.

The Australia Telescope is founded; it has seven 22-m/72-ft dishes and one 64-m/210-ft dish, which are spread throughout New South Wales, Australia, making it one of the largest radio telescopes in the world.

August 8. The European Space Agency launches the *Hipparcos* satellite; it carries two telescopes for measuring the distance of stars.

November 18. The United States launches the *Cosmic Background Explorer* (COBE) satellite to study microwave background radiation, thought to be a vestige of the Big Bang.

1990

January 24. Japan launches *Muses-A*, the first probe to be sent to the moon since 1976; it places a small satellite in lunar orbit on March 19.

February. The U.S. space probe *Voyager 1* (launched September 5, 1977), now near the edge of the solar system, turns and takes the first photograph of the entire solar system from space.

April 24. The U.S. space shuttle *Discovery* places the Hubble Space Telescope in Earth orbit; the main mirror proves to be defective.

June 1. The joint U.S.-German-British X-ray *Röntgensatellite* (ROSAT) is launched.

July. U.S. astronomers Juan Uson, Stephen Boughin, and Jeffrey Kuhn announce the discovery of the largest-known galaxy; more than 1 billion light-years away from Earth, it has a diameter of 5.6 million light-years—almost sixty times that of the Milky Way—and contains about two trillion stars.

August 10. The U.S. *Magellan* radar mapper (launched May 4, 1989) arrives in orbit around Venus; it transmits the most detailed pictures of the planet's surface yet produced.

October 6. NASA and the European Space Agency launch *Ulysses* from the U.S. space shuttle *Discovery* to study the sun.

November. The Keck 1 Telescope on Mauna Kea volcano, Hawaii, is erected; its 10-m/33-ft reflector, composed of thirty-six segments, makes it the largest optical telescope in the world.

U.S. astronomer Mark Showalter discovers an eighteenth moon of Saturn when analyzing pictures transmitted by *Voyager 2*.

1991

The New Technology Telescope, an optical telescope that is part of the European Southern Observatory at La Silla, Chile, comes into operation. Its thin lightweight mirror, 3.5 m/138 in across, is kept in shape by computer-adjustable supports to produce a sharper image than is possible with conventional mirrors.

January. An asteroid 16 km/10 mi in diameter passes between the moon and Earth, scoring a near miss.

April 7. The U.S. space shuttle *Atlantis* launches the *Compton Gamma Ray Observatory* into Earth orbit to study gamma rays and their sources. It weighs 16.7 metric tons/18.4 tons and is the heaviest payload ever carried by a space shuttle.

May 18–26. British chemist Helen Sharman becomes the first Briton to go into space, as a participant in a Soviet space mission launched in *Soyuz TM-12*. She spends six days with Soviet cosmonauts aboard the *Mir* space station.

June 5. The U.S. shuttle *Columbia* launches the *Spacelab Life Sciences-1* laboratory. Astronauts conduct experiments on themselves, rats, and jellyfish polyps.

July 17. The European Space Agency launches its first remote-sensing satellite (*ERS-1*) into polar orbit to monitor Earth's temperature from space.

October 29. The U.S. space probe *Galileo* (launched October 18, 1989) takes the closest ever picture of an asteroid—Gaspra—at a distance of 1,600 km/900 mi.

1992

February 8. The joint NASA-European Space Agency space probe *Ulysses* flies over the north and south poles of Jupiter to enter a trajectory for reaching the south pole of the sun; it transmits data about Jupiter's magnetosphere.

May 5. The National Aeronautics and Space Administration (NASA) launches the new shuttle craft *Endeavour,* named for the eighteenth-century vessel captained by English explorer James Cook.

May 14. Astronauts on the U.S. space shuttle *Endeavour* fit a new motor to the *Intelsat 6* satellite and fire it into a new orbit.

July 10. The European Space Agency's *Giotto* space probe is diverted to encounter Comet Grigg-Skellerup.

September 25. The U.S. launches the *Mars Observer* orbiter, the first U.S. mission to Mars in seventeen years.

November. After 359 years, the Roman Catholic Church accepts that Galileo was right: Earth does go around the sun.

The U.S. *Cosmic Background Explorer* (COBE) satellite detects ripples in the microwave background radiation, thought to originate from the formation of galaxies.

The U.S. space probe *Magellan* maps 99 percent of the surface of Venus to a resolution of 100 m/330 ft.

U.S. astronomers Jeffrey McClintock, Ronald Remillard, and Charles Bailyn identify Nova Muscae as a black hole approximately 18,000 light-years from Earth.

1993

March 24. U.S. astronomers Carolyn and Eugene Shoemaker and David Levy report the first sighting of Comet Shoemaker-Levy 9. The comet was found to consist of at least twenty-one fragments in an unstable orbit around Jupiter.

March. A star in the galaxy M81, about 11 million light-years away from Earth, erupts into a supernova. Archival photographs allow astronomers to study the behavior of the star before it exploded.

August 21. NASA loses contact with its $980 million *Mars Observer.*

August 28. The U.S. space probe *Galileo* discovers the first asteroid moon. About 1.5 km/0.95 mi across and named Dactyl (in 1994), it orbits the asteroid Ida.

December 7. The Hubble Space Telescope (placed in Earth orbit in 1990) is repaired and reboosted into a nearly circular orbit by five U.S. astronauts operating from the U.S. space shuttle *Endeavour*—at a cost of $360 million.

U.S. astronomers identify part of the dark matter in the universe as stray planets and brown dwarfs. Known as MACHOs (massive astrophysical compact halo objects), they may constitute approximately half of the dark matter in the Milky Way's halo.

U.S. astronomers Jane Luu and David Jewitt discover four large ice objects in the solar system beyond Pluto. They are the first members of the Kuiper belt (a ring of small, icy bodies orbiting the sun beyond the planets and thought to be the source of comets) to be observed.

U.S. astronomers Russell Hulse and Joseph Taylor Jr. are jointly awarded the Nobel Prize for Physics for their discovery of a new type of pulsar.

1994

January 8. The Russians launch *Soyuz-TM 18*, with cosmonaut Valeri Polyakov aboard, to their *Mir* space station. Polyakov plans to spend fourteen months at the space station to study the effect on the human body of being in space for the time required to travel to Mars.

January 25–May 10. The United States launches the *Clementine* spacecraft to make scientific observations of the moon and the near-Earth asteroid 1620 Geographos. *Clementine* discovers an enormous crater on the far side of the moon. The South Pole-Aitken crater is 2,500 km/1,563 mi across and 13 km/8 mi deep, making it the largest crater in the solar system discovered so far. It also reveals the possibility of a permanent frozen water-ice deposit in a crater near the south pole of the moon. On May 10, however, a malfunction in one of the onboard computers causes the spacecraft to burn up all of its fuel and go out of control, making the flyby of Geographos impossible.

February. Russian cosmonaut Sergei Krikalev flies with U.S. astronauts Charles Bolden and Kenneth Reightler in the space shuttle *Discovery* (launched February 3, 1994). He is the first cosmonaut to fly on a U.S. mission in space.

July 16–22. Fragments of Comet Shoemaker-Levy 9 collide with Jupiter.

December 6. Pictures taken by the Hubble Space Telescope of galaxies in their infancy are published.

December. The Apollo asteroid (an asteroid with an orbit that crosses that of Earth) 1994 XM1 passes within 100,000 km/60,000 mi of Earth, the closest observed approach of any asteroid.

The closest pulsar to Earth (PSR J0108–1431) is discovered; it is 280 light-years away.

The first discoveries from the Keck Telescopes on Mauna Kea, Hawaii, are published.

1995

January 9. Russian cosmonaut Valeri Polyakov, on board the *Mir* space station, spends his 366th day in space, breaking the record for the longest stay in space. He returns to Earth on March 22, after 439 days.

February 3–11. Eileen Collins pilots *Discovery*, making her the first woman to pilot a U.S. space shuttle.

March 22. Russian cosmonaut Yelena Kondakova, on the *Mir* space station, sets a new record for time spent in space by a woman. She returns to Earth after 170 days.

June 29–July 4. The U.S. space shuttle *Atlantis* docks with the Russian *Mir* space station in the first superpower linkup in space since 1975.

July. The Hubble Space Telescope discovers a 320-km/200-mi yellow spot on the surface of Jupiter's moon Io. Although volcanic in origin, its exact cause is unknown.

July. The U.S. space probe *Galileo* enters an unexpected dust storm 55 million km/34 million mi from Jupiter. It detects twenty thousand particles a day, whereas previously the maximum detected was two hundred.

July 23. Two independent observers, the U.S. astronomers Alan Hale and Thomas Bopp discover Comet Hale-Bopp. It is the brightest periodic comet, and its icy core is estimated to be 40 km/25 mi wide.

November 17. Europe launches the *Infrared Space Observatory* (ISO) to discover brown dwarfs (cool masses of gas smaller than the sun), which make up much of the dark matter of the galaxy.

December 12. The United States and Europe launch the *Solar and Heliospheric Observatory* (SOHO). Its main purpose is to study the sun's internal structure and the solar wind of atomic particles streaming toward Earth from the sun.

The first comet-size objects in the Kuiper belt are discovered; previously, the only objects found had diameters of at least 100 km/62 mi, whereas comets generally have diameters of less than 10 km/6.2 mi.

The largest (3 m/9.8 ft across) liquid-mirror telescope (a reflecting telescope constructed with a rotating mercury mirror) is completed for NASA's (National Aeronautics and Space Administration) Orbital Debris Observatory in New Mexico.

U.S. astronomers discover the first brown dwarf, an object larger than a planet but not massive enough to ignite into a star, in the constellation Lepus. It is twenty to forty times as massive as Jupiter. Four other brown dwarfs are discovered in 1996.

U.S. astronomers discover water—in the form of superheated steam—in the sun in two sunspots in which the temperature is only 3,000°C/5,400°F.

1996

January 30. Comet Hyakutake is discovered by (and named for) Japanese amateur astronomer Yuji Hyakutake.

January. At a meeting of the American Astronomical Society, U.S. astronomers announce the discovery of three new planets orbiting stars, all within 50 light-years of Earth. By July 1996, the total number of new planets discovered since October 1995 rises to ten.

January. The U.S. space probe *Galileo* (launched October 18, 1989) shows less helium on the planet Jupiter than expected. The ratio of helium to hydrogen is similar to that of the sun, suggesting that the composition of Jupiter has remained unchanged since its formation. The probe also records 700-kph/435-mph winds below one of the cloud layers, suggesting internal heating.

February 17. The United States launches the *Near Earth Asteroid Rendez-vous* (NEAR) spacecraft toward the asteroid Eros. It is to go into orbit around Eros and study its size, shape, mass, magnetic field, composition, and surface and internal structure.

February 26. A $442-million U.S. satellite is lost in space when its 20-km-/12-mi-long tether, which links the satellite to the space shuttle *Columbia*, snaps. The tethered satellite was designed to generate electricity as it was pulled across Earth's magnetic field by the shuttle.

March 24. Comet Hyakutake makes its closest approach, passing within 15.4 million km/9.5 million mi of Earth. It is the brightest comet for decades, with a tail extending over 12° of the sky.

March 25. The Russian space station *Mir* and the U.S. space shuttle *Atlantis* dock. U.S. astronaut Shannon Lucid begins a five-month stay aboard *Mir*.

March 27. The joint U.S.-British-German *Röntgensatellite* (ROSAT) X-ray-astronomy satellite (launched June 1, 1990) records the emission of X-rays from Comet Hyakutake, which are usually associated with a much hotter source.

March. The U.S. *Near Earth Asteroid Tracking* (NEAT) system, in its first full month in operation, detects about two hundred new asteroids.

March. The U.S. space probe *Galileo* detects a magnetic field around Jupiter's moon Ganymede, suggesting that it has a molten core. It also detects molecules containing both carbon and nitrogen, suggesting that life may have existed on the moon.

June 4. The European Space Agency's Arianespace launches the new *Ariane 5* rocket from French Guiana, after ten years of development work. It veers off course immediately after takeoff on its maiden flight and disintegrates (it is blown up for safety reasons), setting the European space program back years.

July 2. The U.S. aerospace company Lockheed Martin unveils plans for the X-33, a $1 billion wedge-shaped rocket ship. Called the *Venture Star,* it will be built and operated by Lockheed Martin and is scheduled to replace the U.S. space-shuttle fleet by the year 2012.

August 6. NASA scientists' report on a meteorite discovered in Antarctica 1984 states that it is fifteen million years old and is believed to have been ejected from Mars millions of years ago, hitting Antarctica twelve thousand years ago. The meteorite is found to contain fossil microorganisms, suggesting that life once existed on Mars. Critics charge that the meteorite was contaminated.

August. Japan launches *Advanced Earth Observing Satellite* (ADEOS) to gather data on climate change, the environment, and Earth and ocean processes.

August 13. NASA (National Aeronautics and Space Administration) scientists report that new images of Europa, one of Jupiter's moons, taken by the U.S. spacecraft *Galileo* show that icy floes on its surface may contain evidence of life.

September 26. U.S. astronaut Shannon Lucid ends her 188-day stay in space. It is the longest stay for a U.S. astronaut and the longest for a woman. She has spent most of the time aboard the Russian space station *Mir.*

October 3. British scientists announce meteorite evidence of life on Mars, supporting claims made by NASA (National Aeronautics and Space Administration) and U.S. scientists in August.

November 7. The United States launches the *Mars Global Surveyor* orbiting spacecraft to study the magnetic field, climate, and composition of the atmosphere of Mars.

November 16. Russia launches the *Mars-96* spacecraft from a site in Kazakhstan, but the booster rockets fail to fire, and it falls back to Earth and crashes into the Pacific two days later.

November 30. The 5-km-/13-mi-long close-approach asteroid Toutates passes within 5 million km/3 million mi of Earth. Traveling at 140,000 kph/85,0000 mph, it comes close to Earth every four or five years.

December 3. U.S. astronomer Anthony Cook, using data from the satellite *Clementine,* announces the discovery of a frozen lake at the bottom of a crater on the dark side of the moon. It would be important for a future moon colony.

December 4. The United States launches the *Mars Pathfinder.* Its main goal is to demonstrate the feasibility of low-cost landings on, and exploration of, Mars. The spacecraft carries a roving machine to explore the surface.

December 6. Cosmonauts aboard the Russian *Mir* spaceship successfully harvest a small wheat crop, the first plants to be successfully cultivated from seed in space.

December 9. The U.S. *Galileo* spacecraft flies within 692 km/430 mi of the cracked and icy surface of Jupiter's moon Europa.

Dutch astronomers from the Leiden Observatory in the Netherlands and U.S. astronomers from Johns Hopkins University in Baltimore, Maryland, using data from the Hubble Space Telescope, discover a black hole in the galaxy in the constellation Virgo.

Based on data received from the U.S. space probe *Galileo,* U.S. astronomers conclude that Jupiter's moon Io has a metallic core. They also detect a 10-megawatt beam of electrons flowing between Jupiter and Io.

Two new objects, each about 500 km/300 mi in diameter, are discovered in the Kuiper belt, four to six times farther from the sun than Neptune.

1997

January. The U.S. Universal Lunarian Society begins offering one-acre sites in the lunar crater Copernicus for $50, with the idea of colonizing the moon. The lunar colony, to be called the City of Lunaria, will consist of sixty-one zones, each 1.6 km/1.0 mile in diameter, covered by elliptical domes containing a controlled atmosphere.

February 27. Canadian astronomer David Gray reports that the star 51 Pegasi, thought to have a planet orbiting it, pulsates in precisely the way needed to mimic the signature of a planet in orbit around it. It casts doubt over the presence of other extrasolar planets discovered in the previous eighteen months.

February 28. The Italian-Dutch satellite *BeppoSAX* (launched April 30, 1996) observes the first visible-light image of a cosmic gamma-ray burst (GRB)—powerful flashes of gamma rays that occur daily, and randomly, and that outshine all other gamma rays combined. The bursts release more energy in ten seconds than the sun will emit in its entire ten-billion-year lifetime, yet, no source has ever been observed. Dutch astronomer Jan van Paradijs and his Italian-Dutch team observe a light source in a distant galaxy that quickly fades after the burst. The bursts were previously thought to be relatively nearby in space.

February. The U.S. space probe *Galileo* (launched October 18, 1989) begins a flyby of Jupiter's moon Europa. It takes photographs of the moon for a potential future landing site.

February. Two new instruments are added to the Hubble Space Telescope: the Near Infrared Camera and Multi-Object Spectrometer (NICMOS), which will enable Hubble to see things farther away (and, therefore, older) than ever before, and the Space Telescope Imaging Spectrograph, which adds a new dimension to the spectral observation of the Hubble Space Telescope and works thirty times faster than its predecessor.

March 23. Comet Hale-Bopp comes to within 190 million km/118 million mi of Earth, the closest since 2000 B.C. NASA (National Aeronautics and Space Administration) launches rockets to study the comet. Its icy nucleus is estimated to be 40 km/25 mi wide, making it at least ten times larger than that of Comet Hyakutake and twice the size of Halley's comet.

March. U.S. astronomers announce that the space probe *Galileo* has detected molecules containing carbon and nitrogen on Jupiter's moon Callisto, suggesting that life once existed on the satellite.

April 8. British astronomer Alan Penny announces the European Space Agency's Project Darwin, a collection of six infrared telescopes forty times more powerful than the U.S. Hubble Space Telescope. Planned to be launched in the year 2015, Darwin's telescopes would have to be stationed between Mars and Jupiter to avoid the light reflected by dust in the inner solar system, and would have to be positioned relative to one another to an accuracy of millionths of a meter. Together they would make a telescope 100 m/328 ft across and would be sensitive enough to detect water or air, and thus life, on a planet 40–50 light-years away from Earth.

April 28. U.S. astronomer William Purcell announces the discovery of a huge stream of antimatter at the heart of the Milky Way galaxy. The jet, the source of which is a mystery, extends for 3,000 light-years above the center of the galaxy.

June 5. U.S. astronomer Jane Luu and colleagues at Harvard University report a new type of object within the solar system—a "worldlet," known by its catalog number 1996TL66, which has a diameter of 500 km/300 mi and which never gets closer to the sun than about 35 astronomical units (AU, where one AU is equivalent to about 149.6 million km/92.96 million mi, the distance between Earth and the sun). They suggest that it represents a new class of object belonging to a population of possibly several thousand orbiting between the Kuiper belt and the Oort cloud.

June 25. During a manually guided docking maneuver, the Russian space station *Mir* collides with its unmanned cargo-supply vessel, causing the space station to lose power and oxygen and to tumble out of control. Repairs are subsequently made.

June 27. The U.S. *Near Earth Asteroid Rendezvous* (NEAR) spacecraft flies within 1,200 km/746 mi of the asteroid Mathilde, taking high-resolution photographs and revealing a 25-km/15.6-mi crater covering the 53-km/33-mi asteroid.

July 4. The U.S. spacecraft *Mars Pathfinder* (launched December 4, 1996) lands on Mars. Two days later, the probe's rover, *Sojourner,* a six-wheeled vehicle that is controlled by an Earth-based operator, begins to explore the area around the spacecraft.

July 10. Japanese astronomer Makoto Hattori and colleagues report the discovery of a knot of mass that they call a "dark cluster." It has the chemical and gravitational properties of a cluster of galaxies but is optically invisible. A new type of cosmic entity, it helps explain how light from a particular quasar has been distorted and challenges the theories of galaxy formation.

August 7. The space shuttle *Discovery* blasts off from Cape Canaveral, Florida, on a twelve-day ozone-research mission to gather environmental data and test new equipment to be used on a future international space station.

August 7. The U.S. *Mars Global Surveyor* spacecraft (launched November 7, 1996) reports the discovery of possible bacterial life on Mars, though this is later refuted.

Mars Pathfinder rover, *Sojourner*. (NASA/JPL)

September 12. The U.S. spacecraft *Mars Global Surveyor* goes into orbit around Mars to conduct a detailed photographic survey of the planet, commencing in March 1998. The spacecraft uses a previously untried technique called "aerobraking" to turn its initially highly elongated orbit into a 400-km/249-mi circular orbit by dipping into the outer atmosphere of the planet.

October 15. NASA and the European Space Agency launch space probe Cassini to the planet Saturn.

"Sakurai's object," a new star named for the amateur Japanese astronomer who discovered it in 1996 in the constellation of Sagittarius, expands since its discovery from an Earth-size hot dwarf, with a surface temperature of 50,000°C/90,032°F, to a bright yellow supergiant about eighty times wider than the sun and no hotter than 6,000°C/10,832°F. Sakurai's object may be a red giant star that had previously shrunk, the contraction of its core triggering nuclear reactions and subsequent reinflation.

A study presented at the Division of Planetary Sciences of the American Astronomical Society says that the moon was created by debris thrown off from Earth after a collision with a planet three times as large as Mars.

The *Solar and Heliospheric Observatory* (SOHO) reveals that Venus's ion-packed tail is 45 million km/28 million mi in length. Discovered in the late 1970s, it stretches away from the sun and is caused by the solar-wind bombardment of the ions in Venus's upper atmosphere.

1998

January 6. The United States launches the *Lunar Prospector* spacecraft to gather information on the moon's resources, structure, and origin.

January 8. U.S. astronomers present evidence at a meeting of the American Astronomical Society in Washington, D.C., that the universe will never stop expanding and that it is some fifteen billion years old, much older than previous estimates.

January 16. The National Aeronautics and Space Administration (NASA) announces that John Glenn, a Democratic U.S. senator from Ohio who in 1962 became the first American in orbit, will be part of the space shuttle

Discovery team on a ten-day mission to study life sciences. Glenn, who will be seventy-seven, will become the oldest space traveler.

January 24. The U.S. space shuttle *Endeavour* successfully docks with the Russian space station *Mir*. U.S. astronaut S. W. Thomas replaces U.S. astronaut David Wolf on the space station.

January 26. Analysis of high-resolution images from the U.S. *Galileo* spacecraft suggests that the icy crust of Europa, Jupiter's fourth-largest moon, may hide a vast ocean that might be warm enough to support life.

January 27. Al Schultz of the Space Science Institute, Baltimore, Maryland, using the Hubble Space Telescope, announces the discovery of a giant planet, larger than the sun, orbiting Proxima Centauri, the closest star to Earth. It is the first planet outside the solar system to be directly observed.

March 5. U.S. scientists announce that the *Lunar Prospector* satellite has detected 11 million metric tons/12 million tons of water on the moon. It is in the form of ice.

March 11. U.S. astronomer Brian Marsden predicts that an asteroid will pass within 48,000 km/30,000 mi of Earth in the year 2028, raising the possibility that it could hit Earth. The following day, however, scientists from the U.S. National Aeronautics and Space Administration (NASA) refute his theory, saying that the asteroid is not likely to come closer than 960,000 km/600,000 mi to Earth.

March 12. Astronomers at Mauna Kea, Hawaii, announce the sighting of a new galaxy, named 0140+326RD1, which is 90 million light-years farther from Earth than the most distant previously known galaxy.

April 6. The U.S. National Aeronautics and Space Administration (NASA) releases a new picture of a rock formation on Mars. A previous picture made the formation look like a face, fueling the theory that the formation was constructed by a Martian civilization. The new picture, which does not look like a face, refutes this theory.

July 4. Japan launches *Planet-B* to study Mars.

October 29–November 7. U.S. astronaut John Glenn, seventy-seven, becomes the oldest space traveler aboard the space shuttle *Discovery*.

November 20. Russia launches the *Zarya* module to begin the International Space Station (ISS) project.

December 3. The space shuttle *Endeavour* is launched on mission STS 88 to dock the U.S. node connection module, *Unity,* to the *Zarya* module in the first ISS assembly mission.

December 10. NASA launches the *Mars Climate Observer,* due to arrive at its destination in September 1999.

1999

NASA launches the *Mars Polar Lander,* due to arrive on Mars in December 1999.

February 7. NASA launches *Stardust* to rendezvous with the comet Wild 2 in 2004 to collect material to be returned to Earth in 2006.

February 20. Russia launches what will be the final crew to the *Mir* space station.

3

Biographical Sketches

Baade, (Wilhelm Heinrich) Walter (1893–1960)

German-born U.S. astronomer who made observations that doubled the distance, scale, and age of the universe. He discovered that stars are in two distinct populations according to their age, known as Population I (the younger) and Population II (the older). Later, he found that Cepheid-variable stars of Population I are brighter than had been supposed and that distances previously calculated from them were wrong.

Baade, born in Shröttinghausen, studied at Universities of Münster and Göttingen in Germany, earning his Ph.D. at the latter institution in 1919. He immigrated to the United States in 1931, working at Mount Wilson Observatory, near Pasadena, California, until 1948 and at Mount Palomar Observatory, near Los Angeles, California, until 1958, when he returned to Germany.

In 1943, during World War II, Baade made use of the blackout darkness to study the Andromeda galaxy and was able to observe, for the first time, some of the stars in the inner regions of that galaxy. He found that the most luminous stars toward the center, which he called Population II stars, are reddish, and he proposed that these have differing structures and origins from the stars in the spiral arms of the galaxy, which he called Population I stars. Population I stars, which are bluish, are young and formed from the dusty material of the spiral arms—hydrogen, helium, and heavier elements; Population II stars, which are reddish, are old, were created near the nucleus of the galaxy, and contain fewer heavy elements.

Bell-Burnell, (Susan) Jocelyn (1943–)

British astronomer. In 1967, she discovered the first pulsar (rapidly flashing star) with British radio astronomer Antony Hewish and colleagues at the Mullard Radio Astronomy Observatory in Cambridge, England.

Born Susan Jocelyn Bell in Belfast, Northern Ireland, near enough to the Armagh Observatory to spend much time there as a child, she was educated at Glasgow and Cambridge Universities. It was while she was a research student at Cambridge that the discovery of pulsars was made. Between 1968 and 1982, she conducted research in gamma-ray astronomy at the University of Southampton, England, and in X-ray astronomy at the Mullard Space Science Laboratory, University College, London. Subsequently, she worked on infrared and optical astronomy at the Royal Observatory, Edinburgh; since 1991, she has been professor of physics at the Open University, Milton Keynes, England.

Bell-Burnell (then Bell) spent her first two years in Cambridge building a radio telescope that was specially designed to track quasars. The telescope had the ability to record rapid variations in signals. In 1967, she noticed an unusual signal, which turned out to be composed of a rapid set of pulses that occurred precisely every 1.337 sec. One attempted explanation of this curious phenomenon was that it emanated from an interstellar beacon, so initially it was nicknamed LGM, for Little Green Men.

Within a few months, however, Bell located three other similar sources. They, too, pulsed at an extremely regular rate, but their periods varied over a few fractions of a second, and they all originated from widely spaced locations in our galaxy. Thus, it seemed that a more likely explanation of the signals was that they were being emitted by a special kind of star: a pulsar.

Bondi, Hermann (1919–)

Austrian-born British cosmologist. In 1948, he joined with English astronomer and cosmologist Fred Hoyle and Austrian-born U.S. astronomer Thomas Gold in developing the steady-state theory of cosmology, which suggested that matter is continuously created in the universe.

Bondi was born in Vienna and studied mathematics at Cambridge University, England. It was while he was interned in Britain in 1940 as an "enemy alien" during World War II that he met Gold. Bondi returned to Cambridge University in 1941 and began to do naval-radar work for the British Admiralty in 1942; it was through this work that he met Hoyle. Gold soon joined them, and the three discussed cosmology and related subjects.

In 1954, Bondi took a chair in applied mathematics at King's College, London. He was elected a Fellow of the Royal Society in 1959 and in 1983 became Master of Churchill College, Cambridge.

The steady-state model stimulated much debate for, while its ideas were revolutionary, it was fully compatible with existing knowledge. However, evidence that the universe had once been denser and hotter emerged in the 1950s and 1960s, and the theory was abandoned by most scientists.

Cannon, Annie Jump (1863–1941)

U.S. astronomer who carried out revolutionary work on the classification of stars by examining their spectra, a distinctive pattern produced by each star when its light is passed through a prism. Her system, still used today, arranges spectra into categories according to temperature.

Cannon was born in Dover, Delaware, and studied at Wellesley College in Boston and Radcliffe College in Cambridge, Massachusetts. She spent her career at the Harvard College Observatory in Cambridge as assistant (1896–1911), curator of astronomical photographs (1911–1938), and astronomer and curator (1938–1940).

Studying photographs of the spectra of stars, Cannon discovered three hundred new variable stars. In 1901, she published a catalog of the spectra of more than a thousand stars, using her new classification system. She went on to classify the spectra of more than 300,000 stars. Most of this work was published in a ten-volume set, which was completed in 1924.

Dr. Annie Cannon works in her Boston office as curator of the Harvard Observatory Record, 30 April 1930. She is the only woman to hold an honorary degree of Doctor of Science from Oxford University. (Corbis-Bettmann)

It described almost all stars with magnitudes greater than nine. Her later work included classification of the spectra of even fainter stars.

Cannon's system replaced one developed in 1890 by U.S. astronomer Williamina Fleming, which established a system for classifying stellar spectra into categories labeled alphabetically (A–Q). Cannon's categories, based on temperature, are O, B, A, F, G, K, M, R, N, and S. O-type stars are the hottest, with surface temperatures ranging from 25,000 to 50,000 K (24,700°C/44,450°F to 49,700°C/89,500°F). According to her system, stars in the O, B, A group are white or bluish; those in the F, G group, yellow; those in the K group, orange; and those in the M, R, N, S group, red. Our sun is yellow, so its spectrum places it in the G group.

Chandrasekhar, Subrahmanyan (1910–1995)

Indian-born U.S. astrophysicist who made pioneering studies of the structure and evolution of stars. The Chandrasekhar limit is the maximum mass of a white dwarf before it turns into a neutron star. He and U.S. astrophysicist William Alfred Fowler won the Nobel Prize for Physics in 1983 for their work on the life cycle of stars and the origin of chemical elements.

Chandrasekhar also investigated the transfer of energy in stellar atmospheres by radiation and convection and the polarization of light emitted from particular stars.

Chandrasekhar was born in Lahore (now in Pakistan) and studied at Presidency College, University of Madras, India, graduating in 1930 with a B.A., and at Cambridge University, England, where he gained a doctorate in theoretical physics. He joined the staff of the University of Chicago, Illinois, in 1936, becoming a professor there in 1952.

Chandrasekhar explained the evolution of white dwarfs in his *Introduction to the Study of Stellar Structure* (1939). He calculated that stellar masses below 1.4 times that of the sun would form stable white dwarfs, but those above this limit would not evolve into white dwarfs; the limit is now believed to be about 1.2 solar masses. Stars with masses above the Chandrasekhar limit are likely to explode into supernovae; the mass remaining after the explosion may form a white dwarf if the conditions are suitable, but it is more likely to form a neutron star.

Eddington, Arthur Stanley (1882–1944)

British astrophysicist who studied the motions, equilibrium, luminosity, and atomic structure of the stars. In 1919, his observation of stars during a solar eclipse confirmed German-born U.S. physicist Albert Einstein's prediction that light is bent when passing near the sun, in accordance with the general theory of relativity. In *The Expanding Universe* (1933), Eddington expressed the theory that, in the spherical universe, the outer galaxies or, spiral nebulae, are receding from one another.

Eddington discovered the fundamental role of radiation pressure in the maintenance of stellar equilibrium, explained the method by which the energy of a star moves from its interior to its exterior, and, in *Internal Constitution of the Stars* (1926), showed that the luminosity of a star depends almost exclusively on its mass—a discovery that caused a complete revision of contemporary ideas on stellar evolution.

Eddington was born in Kendal, Cumbria, and educated at Owen's College, Manchester, and Cambridge University, all in England. He was chief assistant at the Royal Observatory, Greenwich, England (1906–1913), led an expedition to Malta in 1909 to determine the longitude of the geodetic station there, and went to Brazil in 1912 as the leader of an expedition to observe a total solar eclipse. At Greenwich, his work was mainly concerned with the analysis of stellar proper motions—the apparent movement of a star in the celestial sphere, determined by the actual movement of the star in space and the movement of the sun—which were then just beginning to be available in reasonable numbers.

In 1913, Eddington returned to Cambridge as professor of astronomy and was director of the university's observatory from 1914. There, his first care was to complete the meridian observations begun by his predecessor. His first book, *Stellar Movements and the Structure of the Universe* (1914), introduced the subject of stellar dynamics. He also became interested in Einstein's theories of relativity and, realizing that the total solar eclipse on May 29, 1919, would be an opportunity to test one of its crucial predictions, accompanied the expedition that successfully confirmed it. His work on the theory of relativity appears in *Report on the Relativity Theory of Gravitation* (1918), *The Mathematical Theory of Relativity* (1923), and, in a slightly more popular form, *Space, Time, and Gravitation* (1920). From 1930 onward, he worked on the relationship between the theory of relativity and quantum theory, in *Relativity Theory of Protons and Electrons* (1936), *Philosophy of Physical Science* (1939), and the posthumous *Fundamental Theory* (1948).

Eddington was elected Fellow of the Royal Society in 1914 and was president of the International Astronomical Union (1938–1944).

Fleming, Williamina Paton Stevens (1857–1911)

Scottish-born U.S. astronomer. She was an assistant to Edward Pickering, director of the Harvard College Observatory in Cambridge, Massachusetts, with whom she compiled the first general catalog classifying stellar spectra, the distinctive pattern produced by each star when its light is passed through a prism.

Fleming was born in Dundee, Scotland, and worked as a schoolteacher before immigrating to the United States with her husband, James Fleming, in 1878. Pickering employed her in 1879, initially as a "computer" and copy editor. In 1898, Fleming was appointed curator of astronomical photographs at the observatory.

Fleming developed her spectral-classification system by studying photographs of the spectra obtained by using prisms placed in front of the objectives of telescopes (that is, the lenses and/or mirrors that collect the light and produce the primary image). In the course of her analysis, she discovered fifty-nine nebulae, more than three hundred variable stars, and ten novae. Her system was adopted in the *Draper Catalogs* (1890), which listed 10,351 stellar spectra in seventeen categories (A-Q). This system was to be superseded by the work of U.S. astronomer Annie Jump Cannon at the same observatory in the early decades of the twentieth century.

Fowler, William Alfred (1911–1995)

U.S. astrophysicist. In 1983, he and Indian-born U.S. astrophysicist Subrahmanyan Chandrasekhar were awarded the Nobel Prize for Physics for their work on the life cycle of stars and the origin of chemical elements.

Fowler was born in Pittsburgh, Pennsylvania, and studied physics at Ohio State University, Columbus, obtaining a bachelor's degree in physics in 1933, and then at the California Institute of Technology (Caltech), Pasadena, where he first gained a Ph.D., and then became a Research Fellow in 1936. He remained at Caltech his entire career, ultimately rising to the position of Instructor Professor of Physics in 1970.

Fowler concentrated on research into the abundance of helium in the universe. The helium abundance was first defined as the result of the "hot Big Bang" theory proposed by U.S. physicist Ralph Alpher, German-born U.S. physicist Hans Bethe, and Russian-born U.S. cosmologist George Gamow in 1948. In its original form, the Big Bang theory accounted for the creation of only the lightest elements, hydrogen and helium. In their classic 1957 paper, "B^{2}FH," Fowler and English astronomers Fred Hoyle and Margaret and Geoffrey Burbidge described how, in a star like the sun, two hydrogen nuclei, or protons, combine to create the next-heavier

William A. Fowler of the California Institute of Technology shared the 1983
Nobel Prize for Physics with Subrahmanyan Chandrasekhar of the University of
Chicago, for their collaborative research into how stars are born. Fowler is shown here
at the Yerkes Observatory in Williams Bay, Wisconsin, attending a workshop, 1983.
(Corbis-Bettmann)

element, helium, thus generating energy. Over time, more and heavier elements are produced until, after millions of years, the star finally explodes into a supernova, scattering its material across the universe.

Fowler and Hoyle published an even more complete exposition of stellar nuclear synthesis in 1965 and completed the work two years later with Robert Wagoner. Taking into account all of the reactions that can occur between the light elements, and considering the buildup of heavier elements, they were able to calculate helium abundance in the universe to an accuracy of 1 percent.

Gamow, George (Georgi Antonovich) (1904–1968)

Russian-born U.S. cosmologist, nuclear physicist, and popularizer of science. His work in astrophysics included a study of the structure and evolution of stars and the creation of the elements. He explained how the collision of nuclei in the solar interior could produce the nuclear reactions that power the sun. With the "hot Big Bang" theory, which he coproposed in 1948, he indicated the origin of the universe.

Gamow predicted that the electromagnetic radiation left over from the universe's formation should, after having cooled down during the subsequent expansion of the universe, manifest itself as a microwave cosmic background radiation. He also made an important contribution to the understanding of protein synthesis.

Gamow was born in Odessa (now in Ukraine) and studied at the University of Leningrad, St. Petersburg, where he took courses in optics and later cosmology, gaining his Ph.D. in 1928. He then went to the University of Göttingen, Germany, where his career as a nuclear physicist rally began. His work impressed Danish physicist Niels Bohr, who invited him to the Institute of Theoretical Physics in Copenhagen, Denmark. He then worked with British physicist Ernest Rutherford at the Cavendish Laboratory, Cambridge, England (1929–1930). From 1931 to 1933, he was Master of Research at the Academy of Science, Leningrad, and then defected to the United States, where he held the chair of physics at George Washington University in Washington, D.C. (1934–1956), and was Professor of Physics at the University of Colorado (1956–1968). In the late 1940s, he worked on the hydrogen bomb at Los Alamos, New Mexico.

Gamow's model of alpha decay, presented in 1928, represented the first application of quantum mechanics to the study of nuclear structure. Later, he described beta decay.

With U.S. physicist Ralph Alpher and German-born U.S. physicist Hans Bethe, he investigated the possibility that heavy elements could have been produced by a sequence of neutron-capture thermonuclear reactions. They published a paper in 1948, the premise of which became known as the Alpher-Bethe-Gamow (or alpha-beta-gamma) hypothesis, describing the "hot Big Bang."

Gamow also contributed to the solution of the genetic code. The double-helix model for the structure of DNA (deoxyribonucleic acid) involves four types of nucleotides (complex organic compounds). Gamow realized that if three nucleotides were used at a time, the possible combinations could easily code for the different amino acids of which all proteins are constructed. Gamow's theory was found to be correct in 1961.

Giacconi, Riccardo (1931–)

Italian-born U.S. physicist. He is the head of a team whose work has been fundamental in the development of X-ray astronomy. In 1962, a rocket Giacconi and his group sent up to observe secondary spectral emission from the moon detected strong X rays from a source evidently located

Riccardo Giacconi explains a diagram of two previously unknown sources of space radiation that were found in the Milky Way accidentally in 1962 during moon research by the air force. Dr. Giacconi is one member of a five-man team working for the air force. (Corbis-Bettmann)

outside the solar system. X-ray research has since led to the discovery of many types of stellar and interstellar material. Giacconi and his team developed a telescope capable of producing X-ray images. In 1970, they launched a satellite called *Uhuru,* which was devoted entirely to X-ray astronomy—the detection of stellar and interstellar material whose emissions lie in the X-ray band of the electromagnetic spectrum.

Giacconi was born in Genoa and obtained his doctorate in 1954 from the University of Milan. He immigrated to the United States in 1956, becoming Research Associate first at the University of Indiana at Bloomington and then at Princeton. In 1959, he joined American Scientific and Engineering, Inc., as senior scientist, rising to executive vice president in 1969. In 1973, he was made professor of astronomy at Harvard University, Cambridge, Massachusetts. Later, he became director of the European Southern Observatory.

Hale, George Ellery (1868–1938)

U.S. astronomer. More than any other person, it is he who was responsible for the development of observational astrophysics in the United States. He made pioneering studies of the sun and founded three major observatories. In 1889, he invented the spectroheliograph, a device for photographing the sun at particular wavelengths. In 1917, he established a 2.5-m/100-in reflector on Mount Wilson, near Los Angeles, California; it was the world's largest telescope until it was superseded in 1948 by the 5-m/200-in reflector on Mount Palomar, near Pasadena, California, which Hale had planned just before he died. He also founded the Yerkes Observatory in Williams Bay, Wisconsin, in 1897, with the largest refractor, at a diameter of 102 cm/40 in, ever built at that time.

Hale was born in Chicago, Illinois, and studied first at the city's Adam Academy. He then studied physics, chemistry, and mathematics at the Massachusetts Institute of Technology, Cambridge. The son of a wealthy father, he was able, as soon as he had graduated in 1890, to found a private observatory at Kenwood, near Chicago. It was there that he developed the spectroheliograph and started the studies of the sun that were to remain his main research interest. In 1892, he became a professor of astrophysics at the University of Chicago, and, in 1895, he founded the *Astrophysical Journal,* still the most important international periodical in the field.

Hale persuaded Chicago industrialist C. T. Yerkes to fund construction of an observatory, which was named for its benefactor. Hale worked at the Yerkes Observatory from its founding in 1897 until 1904, when he went to California in search of better observing conditions. There he organized, and became first director of, the Mount Wilson Solar Observatory, which, in addition to its two special solar-tower telescopes, quickly acquired a 1.5-m/60-in and a 2.5-m/100-in reflector. Hale retired from the Mount Wilson Observatory in 1923 but continued his solar observations at his own private observatory in Pasadena, where the Mount Wilson Observatory offices were situated and where he had been partly responsible for developing California Institute of Technology from Throop Polytechnic Institute in Pasadena.

In 1928, Hale initiated the project that resulted in the 5-m/200-in reflector on Mount Palomar. When this instrument was dedicated in 1948, it was formally named the Hale Telescope. Some years later, the Mount Wilson and Mount Palomar Observatory complex was renamed the Hale Observatories. Hale was the author of many scientific papers and of several semipopular books on astronomy.

His work on solar spectra was the stimulus for his construction of a number of specially designed telescopes. By studying the split spectral lines of sunspots, he showed the presence of very strong magnetic fields— the first discovery of a magnetic field outside Earth. In 1919, he showed that these magnetic fields reverse polarity twice every twenty-two to twenty-three years.

Hoyle, Frederick (1915–)

English astronomer, cosmologist, and writer whose astronomical research has dealt mainly with the internal structure and evolution of the stars. In 1948, with Austrian-born British cosmologist Hermann Bondi and Austrian-born U.S. astronomer Thomas Gold, he developed the steady-state theory of the universe, which suggested that matter is continuously created in the universe.

In 1957, with U.S. astrophysicist William Fowler, he showed that chemical elements heavier than hydrogen and helium may be built up by nuclear reactions inside stars. Fowler and Hoyle proposed that all of the elements may be synthesized from hydrogen by successive fusions. When the gas cloud reaches extremely high temperatures, the hydrogen has turned to helium and neon, whose nuclei interact, releasing particles that unite to build up nuclei of new elements.

Hoyle was born in Bingley, Yorkshire, and studied at the University of Cambridge, England. He spent his academic career at Cambridge, where he was a lecturer and later professor of astronomy (1958–1973), and at Mount Palomar Observatory, near Pasadena, California (1956–1966). He was director of the Cambridge Institute of Theoretical Astronomy (1967–

1973), which he was instrumental in founding. He has also held various visiting professorships in England and the United States. He was elected Fellow of the Royal Society in 1957.

Hubble, Edwin Powell (1889–1953)

U.S. astronomer who discovered the existence of galaxies outside our own and classified them according to their shape. His theory that the universe is expanding is now generally accepted.

His data on the speed at which galaxies were receding (based on their red shifts) were used to determine the portion of the universe that we can ever come to know, the radius of which is called the Hubble radius. Beyond this limit, any matter will be traveling at the speed of light, so communication with it will never be possible. The ratio of the velocity of galactic recession to distance has been named the Hubble constant.

In 1924, Hubble discovered Cepheid-variable stars in the Andromeda galaxy, proving it to lie far beyond our own galaxy. In 1925, he introduced the classification of galaxies as spirals, barred spirals, and ellipticals. In 1929, he announced Hubble's Law, stating that the galaxies are moving apart at a rate that increases with their distance from each other.

Hubble was born in Marshfield, Missouri, and graduated from the University of Chicago in 1910. He then studied at Oxford University, England, from which he gained a degree in law in 1912. He was also an athlete and a heavyweight boxer. He returned to the United States in 1913 and briefly practiced law in Louisville, Kentucky, but his real interest was in astronomy. In 1914, he began graduate studies in astronomy at the Yerkes Observatory in Williams Bay, Wisconsin, from which he was awarded a Ph.D. in 1917 for a thesis on the photography of faint nebulae.

After serving with the American Expeditionary Forces in World War I, he joined the staff at Mount Wilson Observatory, near Pasadena, California, in 1919 and carried out research on galactic and extragalactic nebulae. During World War II, he was chief of ballistics and director of the Supersonic Wind Tunnel Laboratory at the Aberdeen Proving Ground, Maryland. Afterward, he returned to what was to become the Mount Wilson and Palomar Observatories and was one of the first to use the new 5-m/200-in Hale telescope that was opened in 1948.

Nearly all of Hubble's work related to nebulae, which he was the first to show were extragalactic, that is, outside our own galaxy. It has been said that Hubble opened up the observable region of the universe in the same way that Italian astronomer and physicist Galileo opened up the solar system in the seventeenth century and that British astronomers William

Dr. Edwin P. Hubble, dean of the American astronomers and in charge of the program, is shown putting the unique, 48-inch, Schmidt Photographic Telescope in California through its final series of rehearsal for the National Geographic Society-Palomar Observatory sky survey, to be launched on 1 July 1949. The purpose of the program is to provide the world with the first definitive map of the stars. (Corbis-Bettmann)

and John Herschel opened the Milky Way in the eighteenth and nineteenth centuries. He gave an account of some of his researches in *The Realm of the Nebulae* (1936).

Jansky, Karl Guthe (1905–1950)

U.S. radio engineer who, in 1932, discovered that the Milky Way galaxy emanates radio waves; he did not follow up his discovery, but it marked the birth of radio astronomy.

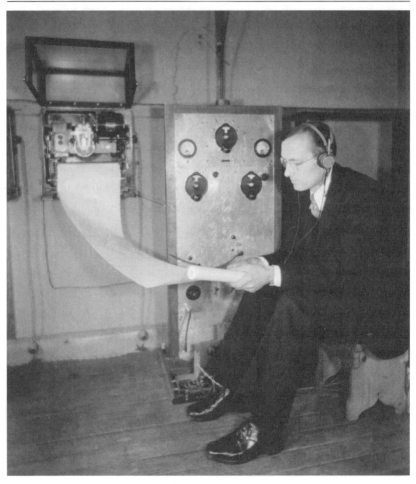

Karl Gunthe Jansky sits beside an instrument used to detect radio waves from the Milky Way at the Bell Telephone Laboratories in New Jersey, 1933. Jansky's discovery of short wave radio waves emanating from the Milky Way led to the inauguration of the science of radio astronomy. (Corbis-Bettmann)

Jansky was born in Norman, Oklahoma, and graduated in physics from the University of Wisconsin. In 1928, he joined the Bell Telephone Laboratories in New Jersey, where he investigated causes of static that created interference on radio-telephone calls.

Jansky noticed that the background hiss on a loudspeaker attached to his specially built receiver-and-antenna system reached a maximum intensity every twenty-four hours. It seemed to move steadily with the sun but gained on the sun by four minutes per day. This amount of time correlates with the difference of apparent motion, as seen on Earth, between the sun and the stars, so Jansky surmised that the source must lie beyond the solar system. By 1932, he had concluded that the source lay in the direction of Sagittarius—the center of our galaxy.

Kuiper, Gerard Peter (1905–1973)

Dutch-born U.S. astronomer who made extensive studies of the solar system. His discoveries included the atmosphere of the planet Mars and that of Titan, the largest moon of the planet Saturn.

Kuiper was adviser to many NASA (National Aeronautics and Space Administration) exploratory missions and pioneered the use of telescopes on high-flying aircraft.

Kuiper was born in Harenkarspel and educated at the University of Leiden. In 1933, he immigrated to the United States, where he joined the staff of the Yerkes Observatory in Williams Bay, Wisconsin, which is affiliated with the University of Chicago, and was its director (1947–1949 and 1957–1960). From 1960 until his death in 1973, he held a similar position at the Lunar and Planetary Laboratory at the University of Arizona.

In 1948, Kuiper correctly predicted that carbon dioxide was one of the chief constituents of the Martian atmosphere. He discovered the fifth moon of Uranus, which he called Miranda, also in 1948; and, in 1949, he discovered the second moon of Neptune, Nereid. Kuiper's spectroscopic studies of Uranus and Neptune led to the discovery of features subsequently named Kuiper bands, which indicate the presence of methane on the observed spectrum.

Lovell, (Alfred Charles) Bernard (1913–)

British radio astronomer and director of the Jodrell Bank Experimental Station (now Nuffield Radio Astronomy Laboratories), Cheshire, England (1951–1981).

During World War II, Lovell worked on developing a radar system to improve the aim of bombers in night raids. After the war, he showed that radar could be a useful tool in astronomy and lobbied for the creation of a radio-astronomy station. Jodrell Bank was built at Cheshire, England (1951–1957). Several large radio telescopes were constructed, including a 76-m/250-ft instrument. Although its high cost was criticized, its public success after tracking the Soviet satellite *Sputnik 1* in 1957 assured its future.

Lovell was born at Oldland Common, Gloucestershire, England, and studied at the University of Bristol, England, where he graduated in physics in 1933. He spent his academic career at the University of Manchester, where he worked on cosmic rays (1936–1939) and, after World War II, as a lecturer in physics. In 1950, he discovered that galactic radio sources emitted at a constant wavelength, and that the fluctuations (scintillation) recorded on Earth's surface were introduced only as the radio waves met and crossed the ionosphere, a layer of the Earth's outer atmosphere. In 1951, he became the first professor of radio astronomy at the University of Manchester. He was made Fellow of the Royal Society in 1955, received a knighthood in 1961, and was president of the Royal Astronomical Society (1969–1971).

Lowell, Percival (1855–1916)

U.S. astronomer who predicted the existence of a planet beyond Neptune, starting the search that led to the discovery of Pluto in 1930. In 1894, he founded the Lowell Observatory in Flagstaff, Arizona, where he reported seeing "canals" (now known to be optical effects and natural formations) on the surface of Mars.

Lowell was born in Boston, Massachusetts, and graduated in mathematics from Harvard University, Cambridge, Massachusetts, in 1876. He spent sixteen years in business and diplomacy, mainly in the Far East, before taking up astronomy and founding the Lowell Observatory. He was made a professor of astronomy at the Massachusetts Institute of Technology, also in Cambridge, in 1902.

Influenced strongly by the work of Italian astronomer Giovanni Schiaparelli, Lowell set up his observatory at Flagstaff originally with the sole intention of confirming the presence of advanced life-forms on Mars. He thought he could make out a complex and regular network of canals and regular seasonal variations that, to him, indicated agricultural activity. He led an expedition to the Chilean Andes in 1907 that produced the first high-quality photographs of the planet.

Maury, Antonia Caetana de Paiva Pereira (1866–1952)

U.S. expert in stellar spectroscopy who specialized in the detection of binary stars. She also formulated a classification system to categorize the appearance of spectral lines.

Maury was born in Cold Spring-on-Hudson, New York, and educated at Vassar College in Poughkeepsie, New York. She became an assistant at the Harvard College Observatory, Cambridge, Massachusetts, under the direction of U.S. astronomer Edward Pickering, even before she graduated in 1887. Later, she lectured in astronomy in various U.S. cities and taught privately. In 1908, she returned to Harvard College Observatory to study the complex spectrum of the binary star Beta Lyrae.

Having assisted Pickering in establishing that the star Mizar (Zeta Ursae Majoris) is a binary star, with two distinct spectra, Maury was the first to calculate the 104-day period of this star. In 1889, she discovered a second such star, Beta Aurigae, and established that it has a period of only four days.

Studying the spectra of bright stars, Maury found three major divisions among spectra, depending upon the width and distinctness of the spectral lines. In 1896, she published her new classification scheme for spectral lines, based on the examination of nearly five thousand photographs and covering nearly seven hundred bright stars in the northern sky.

Oort, Jan Hendrik (1900–1992)

Dutch astronomer. In 1927, he calculated the mass and size of our galaxy, the Milky Way, and the sun's distance from its center, from the observed movements of stars around the galaxy's center. In 1950, Oort proposed that comets exist in a vast swarm, now called the Oort cloud, at the edge of the solar system.

In 1944, Oort's student Hendrik van de Hulst (1918–) calculated that hydrogen in space would emit radio waves at 21-cm/8.3-in wavelength, and in the 1950s Oort's team mapped the spiral structure of the Milky Way from the radio waves given out by interstellar hydrogen.

Oort was born in Franeker, Friesland, and studied at the University of Groningen, receiving his Ph.D. in 1926. Most of his career was spent at the University of Leiden, where he became professor of astronomy in 1935 and director of the observatory (1945–1970).

Oort confirmed calculations made by Swedish astronomer Bertil Lindblad and U.S. astronomer Harlow Shapley concerning the size and shape of our universe and went on to show that the stars in the Milky Way were arranged like planets revolving around a sun, in that the stars nearer the center of the galaxy revolved faster around the center than those farther out did.

He established radio observatories at Dwingeloo and Westerbork, which put the Netherlands in the forefront of radio astronomy. He also played an important part in the national and international organization of astronomy, as a founder of the European Southern Observatory in 1962, and as secretary (1935–1948) and president (1959–1961) of the International Astronomical Union. He was the first recipient of the prestigious Vetlesen Prize, established in 1966 to recognize "achievement in the sciences resulting in a clearer understanding of the earth, its history, or its relation to the universe."

Payne-Gaposchkin, Cecilia Helena (1900–1979)

British-born U.S. astronomer who studied the evolution of stars and the structure of galaxies. Her investigation of stellar atmospheres during the 1920s gave some of the first indications of the overwhelming abundance of the lightest elements (hydrogen and helium) in the galaxy.

Born Cecilia Helena Payne in Wendover, Buckinghamshire, England, she studied at Cambridge University, England (graduating in 1923), and at Harvard College Observatory, Cambridge, Massachusetts, under U.S. astronomer Harlow Shapley. In 1927, she was appointed an astronomer at the observatory. In 1938, she was made Phillips Astronomer at Harvard University before being made professor of astronomy in 1956.

Payne-Gaposchkin employed a variety of spectroscopic techniques in the investigation of the properties and composition of stars, especially variable stars (stars that vary in brightness). She carried out her studies of the Large and Small Magellanic Clouds in collaboration with her husband, Sergei I. Gaposchkin.

Other areas of interest for her included devising methods to determine stellar magnitudes, the position of variable stars on the Hertzsprung-Russell diagram, and novae.

Penzias, Arno Allan (1933–)

German-born U.S. radio engineer who, in 1964, with U.S. radio astronomer Robert Wilson, was the first to detect cosmic background radiation. This radiation had been predicted on the basis of the "hot Big Bang" model of the origin of the universe, initially proposed in 1948. Penzias and Wil-

Arno Penzias and Robert Wilson stand next to the horn antenna they used to discover remnants of cosmic background radiation, the first direct evidence of the Big Bang, AT&T Bell Laboratories, 26 May 1993. (Corbis/Robert Ressmeyer)

son shared the 1978 Nobel Prize for Physics for their work on microwave radiation.

Penzias was born in Munich, Bavaria. His parents left Nazi Germany for the United States, and Penzias studied at the City College of New York, earning his bachelor's degree in physics in 1954, and at Columbia University, where he was awarded his master's degree in 1958 and his doctorate in 1962. In 1961, he joined the staff of the Radio Research Laboratory of the Bell Telephone Laboratories in New Jersey; he was named its director in 1976 and was vice president of research between 1981 and 1995. He became vice president and chief scientist of Bell Labs Innovations in 1996. He has held a series of academic positions at Princeton and Harvard Universities.

In 1963, Penzias and Wilson were assigned by Bell to trace radio noise that was interfering with the development of a communications program involving satellites. They had detected a surprisingly high level of microwave radiation that had no apparent source (that is, it was uniform in all directions). The temperature of this background radiation was 3.5 K (-269.7°C/-453.4°F), later revised to 2.735 K (-270.4°C/-454.7°F)

They took this enigmatic result to U.S. physicist Robert Dicke at Princeton University, Princeton, New Jersey, who had predicted that this sort of radiation should be present in the universe as a residual relic of the intense heat associated with the birth of the universe following the Big Bang. His department was in the process of constructing a radio telescope designed to detect precisely this radiation when Penzias and Wilson presented their data.

Penzias's later work has been concerned with developments in radio astronomy, instrumentation, satellite communications, atmospheric physics, and related matters.

Ryle, Martin (1918–1984)

British radio astronomer. At the Mullard Radio Astronomy Observatory, Cambridge, England, he developed the technique of sky-mapping using aperture synthesis, which involves combining smaller-dish aerials to give the characteristics of one large one. His work on the distribution of radio sources in the universe brought confirmation of the Big Bang theory. He was knighted in 1966, and won, with his coworker, English radio astronomer Antony Hewish, the Nobel Prize for Physics in 1974.

Born in Brighton, Sussex, England, Ryle graduated from Oxford University in 1939. During World War II, he was involved in the development of radar. After the war, he joined the Cavendish Laboratory in Cambridge, and, in 1957, he became the first director of Cambridge University's Mullard Radio Astronomy Observatory. In 1959, he became the first Cambridge professor of radio astronomy, responsible for most of the radio-telescope developments at the university. Ryle became especially interested in isolated radio sources and their cosmological implications. He was elected a Fellow of the Royal Society in 1952 and was Astronomer Royal (1972–1982), an honorary post awarded to an outstanding British astronomer.

Increasingly larger radio telescopes were built at the Cambridge sites, resulting in the *Cambridge Catalog Surveys,* numbered 1C–5C, giving increasingly better maps of radio sources in the northern sky. The 3C survey, published in 1959, is used as a reference by all radio astronomers. The 4C survey cataloged five thousand sources.

The first supersynthesis telescope, in which a fixed aerial maps a band of the sky using solely the rotation of Earth and another aerial maps successive rings out from it concentrically, was built in 1963, and the eight-dish 12.8-m/42-ft-wide Ryle Telescope, in a line 5 km/3 mi long, opened at Mullard in 1972. The programs for which the Ryle Telescope is used include the mapping of extragalactic sources and the study of supernovae and newly born stars. It can provide as sharp a picture as the best ground-based optical telescopes.

Seyfert, Carl Keenan (1911–1960)

U.S. astronomer and astrophysicist who studied the spectra of stars and galaxies and identified and classified the type of galaxy that now bears his name (Seyfert galaxy).

Seyfert was born in Cleveland, Ohio, and studied at Harvard University, Cambridge, Massachusetts, which awarded him a bachelor's degree in 1933, a master's degree in 1935, and a doctorate in 1936. He was director of Barnard Observatory in Oxford, Mississippi (1946–1951), and, from 1951, professor of astronomy at Vanderbilt University, Nashville, Tennessee, and director of the Arthur S. Dyer Observatory there. He was also a member of the National Defense Research Committee.

In 1943, Seyfert's investigations of twelve active spiral galaxies with bright nuclei showed that these galaxies contain hydrogen as well as ionized oxygen, nitrogen, and neon. On the basis of their spectra, Seyfert divided the galaxies into two types, I and II. Seyfert galaxies emit radio waves, infrared energy, X-rays, and nonthermal radiation. The gases at their centers are subject to explosions that cause them to move violently—with speeds of many thousands of kilometers per second relative to the center of the galaxy in the case of type I, and of several hundreds of kilometers per second in the case of type II.

In 1951, Seyfert began a study of the objects now known as Seyfert's Sextet: a group of diverse extragalactic objects, of which five are spiral nebulae and one is an irregular cloud. One member of the group is moving away from the others at a velocity nearly five times that at which the others are receding from each other.

Shapley, Harlow (1885–1972)

U.S. astronomer. He established that our galaxy is much larger than previously thought. His work on the distribution of globular clusters showed that the sun was not at the center of the galaxy, as then assumed, but, rather, two-thirds of the way out to the rim; globular clusters were arranged in a halo around the galaxy.

Shapley was born in Nashville, Missouri, and earned his B.A. in mathematics and physics from the University of Missouri in Columbia, Missouri, in 1910 and his M.A. a year later. He worked at Mount Wilson Observatory, near Los Angeles, California (1914–1921), and then moved to Harvard University, Cambridge, Massachusetts, where he was professor of astronomy and director of Harvard College Observatory (1921–1952). Alleged by U.S. Senator Joseph McCarthy in 1950 to be a Communist, he was interrogated by the U.S. House Un-American Activities Committee (HUAC).

Shapley obtained nearly ten thousand measurements of the sizes of stars in order to analyze ninety eclipsing binaries, that is, binary stars in which the two stars periodically pass in front of each other as seen from Earth. He also showed that Cepheid-variable stars were pulsating single stars, not double stars. He discovered many previously unknown Cepheid variables and devised a statistical procedure to establish the distance and luminosity of a Cepheid variable. Shapley's surveys recorded tens of thousands of galaxies in both the Northern and Southern Hemispheres.

Tombaugh, Clyde William (1906–1997)

U.S. astronomer who discovered the planet Pluto in 1930.

Tombaugh, born in Streator, Illinois, developed a deep fascination for astronomy. His family could not afford to send him to college, so he joined the Lowell Observatory in Flagstaff, Arizona, in 1929 with the intent of learning as much as possible about astronomy. While working there as an assistant, he photographed the sky in search of an undiscovered, but predicted, remote planet.

The new planet would be dim, so each photograph could be expected to show anything between 50,000 and 500,000 stars. And, because of its distance from Earth, any visible motion would be very slight. Tombaugh solved the problem by comparing two photographs of the same part of the sky taken on different days. The photographic plates were focused at a single point and alternately flashed rapidly onto a screen. A planet moving against the background of stars would appear to move back and forth on the screen. Tombaugh found Pluto on February 18, 1930, from plates taken three weeks earlier. He continued his search for new planets across the entire sky; his failure to find any placed strict limits on the possible existence of planets beyond Pluto.

Van Allen, James Alfred (1914–)

U.S. physicist whose instruments aboard the first U.S. satellite, *Explorer 1* in 1958, led to the discovery of the Van Allen belts, two zones of intense radiation around Earth. He also pioneered high-altitude research with rockets after World War II.

Born in Mount Pleasant, Iowa, Van Allen studied at Iowa Wesleyan College, where he received his B.S. degree. He then attended the University of Iowa, Iowa City, where he earned his M.S. in 1936 and his Ph.D. in 1939. He organized and led scientific expeditions to Peru in 1949, the Gulf of Alaska in 1950, Greenland in 1952 and 1957, and Antarctica in

1957 to study cosmic radiation. From 1951, he became professor of physics and head of the same department at the University of Iowa. Since 1972, he has been Carver Professor of Physics there. He participated in Project Matterhorn (1953–1954), which was concerned with the study of controlled thermonuclear reactions, and he was responsible for the instrumentation of the first U.S. satellites.

After the end of World War II, Van Allen began utilizing unused German V2 rockets to measure levels of cosmic radiation in the outer atmosphere, the data being radioed back to Earth. He then conceived of rocket balloons (rockoons), which began to be used in 1952. They consisted of a small rocket that was lifted by means of a balloon into the stratosphere and then fired off.

Whipple, Fred Lawrence (1906–)

U.S. astronomer whose hypothesis in 1949 that the nucleus of a comet is like a dirty snowball was confirmed in 1986 by space-probe studies of Halley's comet.

Whipple was born in Red Oak, Iowa, and received his bachelor's degree from the University of California at Los Angeles in 1927. In 1931, he was appointed to the staff of the Harvard College Observatory in Cam-

Fred Whipple uses a five-hundred-pound snowball covered with dirt to demonstrate the anatomy of a comet's nucleus (1986). (Corbis/Jonathan Blair)

bridge, Massachusetts, becoming a professor there in 1950 until 1977. He was also director of the Smithsonian Astrophysics Observatory in Cambridge, Massachusetts (1955–1973).

In addition to discovering six new comets, Whipple proposed that the nucleus of a comet consisted of a frozen mass of water, ammonia, methane, and other hydrogen compounds together with silicates, dust, and other materials. As the comet's orbit brought it nearer to the sun, solar radiation would cause the frozen material to evaporate, thus producing a large amount of silicate dust that would form the comet's tail.

Whipple also worked on ascertaining cometary orbits and defining the relationship between comets and meteors. In the 1950s, he became active in the program to devise effective means of tracking artificial satellites.

Wilson, Robert Woodrow (1936–)

U.S. radio astronomer who, in 1964, with German-born U.S. radio engineer Arno Penzias, detected cosmic background radiation, which is thought to represent a residue of the primordial Big Bang. He and Penzias shared the 1978 Nobel Prize for Physics for their work on microwave radiation.

Wilson was born in Houston, Texas, and studied at Rice University in that city, earning his bachelor's degree in 1957. He also studied at the California Institute of Technology, Pasadena, where he was awarded a Ph.D. in 1962. In 1963, he joined the Bell Telephone Laboratories in New Jersey; in 1976, he was made head of the Radiophysics Department.

In 1964, Wilson and Penzias tested a radio telescope and receiver system for Bell with the intention of tracking down all possible sources of static that were causing interference in satellite communications. They found a high level of isotropic background radiation at a wavelength of 7.3 cm/2.9 in, with a temperature of 2.735 K (-270.4°C/-454.7°F). This radiation was a hundred times more powerful than any that could be accounted for on the basis of any known sources.

Unable to explain this signal, Wilson and Penzias contacted U.S. physicist Robert Dicke at Princeton University, Princeton, New Jersey, who immediately recognized that their findings confirmed predictions of residual microwave radiation from the beginning of the universe.

Zwicky, Fritz (1898–1974)

Swiss astronomer who, in 1934, predicted the existence of neutron stars. He also discovered eighteen supernovae and determined that cosmic rays originate in them.

Although Zwicky was born in Varna, Bulgaria, his parents were Swiss, and he retained his Swiss nationality throughout his life. He studied at the Federal Institute of Technology in Zürich, Switzerland, gaining a B.A. and then a Ph.D. by 1922. In 1925, he moved to the United States to work as a theoretical physicist at California Institute of Technology, Pasadena, where he was appointed assistant professor of astrophysics in 1927, professor of astrophysics in 1942, and remained until his retirement in 1968.

Zwicky was among the first to suggest that there is a relationship between supernovae and neutron stars. He suggested in the early 1930s that the outer layers of a star that explodes as a supernova leave a core that collapses upon itself as a result of gravitational forces.

Zwicky also observed that most galaxies occur in clusters, each of which contains several thousand galaxies. He made spectroscopic studies of the Virgo and Coma Berenices clusters and calculated that the distribution of galaxies in the Coma Berenices cluster was statistically similar to the distribution of molecules in a gas when its temperature is at equilibrium. Beginning in 1936, he compiled a catalog of galaxies and galaxy clusters in which he listed ten thousand clusters.

4 | Directory of Organizations, Observatories, and Facilities

American Astronomical Society (AAS)

Major professional organization in the United States for astronomers and others interested in astronomy. Publications of the AAS, which was established in 1899, include *Astronomical Journal* and *Astrophysical Journal,* both published by the University of Chicago Press. *Astronomical Journal* appears monthly; *Astrophysical Journal,* three times a month.

Contact Information: 2000 Florida Avenue, Suite 400, Washington, DC 20009; phone: (202) 328-2010; fax: (202) 234-2560; email: aas@aas.org; Web site: http://www.aas.org

Astronomical Journal Editorial Office: Department of Astronomy, University of Washington, Box 351580, Seattle, WA 98195-1580; phone: (206) 685-2150; fax: (206) 685-0403; email: astroj@astro.washington.edu; Web site: http://www.astro.washington.edu/astroj/index.html

Astrophysical Journal Editorial Office: Kitt Peak National Observatory, P.O. Box 26782, 950 N. Cherry Avenue, Tucson, AZ 85726-6732; phone: (520) 318-8214; fax: (520) 318-8183; email: apj@noao.edu; Web site: http://www.noao.edu/apj/apj.htm

Ames Research Center

U.S. facility for the development of new aeronautics technologies. It was founded in 1939 as an aircraft research laboratory; in 1958, it became part of the space-research installation of the National Aeronautics and Space Administration (NASA) at Mountain View, California, for the study of aeronautics and life sciences.

Contact Information: Moffett Field, CA 94035-1000; phone: (650) 604-3574; fax: (650) 604-3445; Web site: http://www.arc.nasa.gov/

Anglo-Australian Observatory

Observatory in New South Wales, Australia, that provides research facilities for the Australian and British astronomical communities. Founded in 1971, the observatory is jointly funded by Britain and Australia. The observatory has two telescopes:

UK Schmidt Telescope (UKST)

A 1.2-m/48-in survey telescope. Commissioned in 1973, it became part of the Anglo-Australian Observatory in 1988. With its wide-angle field of view, it was designed to photograph 6.6-x-6.6° areas of the night sky. Its initial project was the first deep, blue-light photographic survey of the southern skies, which was completed in the 1980s.

Anglo-Australian Telescope (AAT)

A 3.9-m/154-in telescope. Commissioned in 1974, the AAT was one of the last 4-m/158-in equatorially mounted telescopes to be constructed. One of the most commonly used instruments of the AAT are its spectrographs, which split the light from objects into its constituent colors, enabling the resulting spectrum to be analyzed in detail to measure such properties as temperature, chemical composition, velocity, and distance of an object.

Contact Information: P.O. Box 296, Epping, New South Wales 1710, Australia; phone: +61 2 9372 4800; fax: +61 2 9372 4880/4860; email: rec@aaoepp.aao.gov.au; UKST Web site: http://www.aao.gov.au/ukst; AAT Web site: http://www.aao.gov.au/about/aat.htm

Apache Point Observatory

Observatory owned by the Astrophysical Research Consortium, a non-profit consortium of research institutions. The observatory's principal projects are a 3.5-m/138-in telescope and the Sloan Digital Sky Survey. The multipurpose telescope provides astronomers with data to develop and refine theories on how planets, stars, nebulae, and other celestial objects form and evolve. Features of the telescope include remote control and rapid instrument change capability, enabling several astronomers to share the telescope at any one time, using whichever of the various arrays of the telescope's instruments suits their requirements.

Contact Information: 2001 Apache Point Road, P.O. Box 59, Sunspot, NM 88349-0059; phone: (505) 437-6822; fax: (505) 434-5555; email: norm@galileo.apo.nmsu.edu; Web site: http://www.apo.nmsu.edu/

Arecibo Observatory

Observatory situated in Arecibo, Puerto Rico, part of the National Astronomy and Ionosphere Center (NAIC) run by Cornell University and the National Science Foundation. Opened in 1963, it is the site of the world's largest single-dish radio telescope. Its dish-shaped radio mirror is 305 m/1,000 ft in diameter and 51 m/167 ft deep. The surface consists of almost forty thousand perforated aluminum panels, supported by a network of steel cables. It is the largest curved focusing antenna in the world, making it the most sensitive of radio telescopes. The observatory also maintains an Ionospheric Interactions Facility, consisting of thirty-two log-periodic antennae and transmitters capable of concentrating energy in the ionosphere; the effects of these transmissions are studied using the telescope and enable the study of plasma physics.

Contact Information: P.O. Box 995, Arecibo, Puerto Rico 00613; phone: (787) 878-2612; observatory Web site: http://www.naic.edu/about/ao/descrip.htm; telescope Web site: http://www.naic.edu/

Astronomical Society of the Pacific (ASP)

U.S. society formed in 1889 by professional and amateur astronomers who originally joined together to view a total solar eclipse. The ASP's original purpose was to circulate astronomical information; it has since become the largest general astronomy society worldwide, with members from more than seventy countries. It publishes the bimonthly magazine *Mercury*, which is circulated to members of the society.

Contact Information: 390 Ashton Avenue, San Francisco, CA 94112; phone: (415) 337-1100; fax: (415) 337-5205; email: membership@aspsky.org.; Web site: http://www.aspsky.org/

Australia Telescope National Facility

Australian organization that supports and undertakes research in radio astronomy. It is a division of the Commonwealth Scientific and Industrial Research Organization (CSIRO). Based in New South Wales, it operates the Australia Telescope, which consists of eight radio-receiving antennae located at three sites that are used to study objects in the universe ranging from the remains of dead stars to whole galaxies.

Contact Information: P.O. Box 76, Epping, New South Wales 1710, Australia; phone: +61 2 9372-4100; fax: +61 2 9372-4310; email: atnf@atnf.csiro.au; Web site: http://www.atnf.csiro.au/

Calar Alto Observatory

German-Spanish astronomical center based in southern Spain. It operates three telescopes, with apertures of 1.23 m/48.4 in, 2.2 m/87 in, and 3.5 m/138 in, as well as a Schmidt reflector. Also at the site is a 1.5-m/59-in telescope operated by the Observatory of Madrid.

Contact Information: Calar Alto, Sierra de Los Filabres, Andalucia, Spain; fax: (06221) 528-444-518; Web site: http://www.mpia-hd.mpg.de/CAHA/indexA2.htm

Central Bureau for Astronomical Telegrams

Nonprofit U.S. organization responsible for the circulation of information on transient astronomical events through a series of announcements in both printed and electronic form, called the *IAU Circulars*. CBAT operates at the Smithsonian Astrophysical Observatory; its principal funding comes from subscriptions to its various services.

Contact Information: Smithsonian Astrophysical Observatory, 60 Garden Street, Cambridge, MA 02138; phone: (617) 495-7280; fax: (617) 495-7231; email: iausubs@cfa.harvard.edu; Web site: http://cfa-www.harvard.edu/cfa/ps/cbat.html

Cerro Tololo Interamerican Observatory

Chilean observatory, opened in 1974, operated by the Association of Universities for Research into Astronomy (AURA). A complex of astronomical telescopes and instruments, its main instrument is the permanently mounted 4-m/158-in Blanco Telescope. The observatory's facilities are available for approved projects of all qualified astronomers in the Western Hemisphere.

Contact Information: CTIO/AURA Inc., Casilla 603, La Serena, Chile; phone: +56 51 225415; fax: +56 51 205342; email: jhughes@noao.edu; observatory Web site: http://www.ctio.noao/ctio.htm; telescope Web site: http://www.ctio.noao.edu/soar

CHARA Array

U.S. astronomical interferometer project located at the Mount Wilson Observatory, near Pasadena, California. The facility is a project of Georgia State University's Center for High Angular Resolution Astronomy. Construction began in 1996, and first light was expected by the end of the decade; completion of the project was planned for the year 2000.

Contact Information: University Plaza, Atlanta, GA 30303; phone: (404) 651-2000; email: hal@chara.gsu.edu; Web site: http://www.chara.gsu.edu/CHARA/charasummary.html

Effelsberg Radio Observatory

Observatory of the Max Planck Institute for Radio Astronomy, near Bonn, Germany. The Effelsberg 100-m/328-ft radio telescope is the world's largest fully steerable instrument of its kind. It is used solely for research in radio astronomy and for global very long baseline interferometry (VLBI).

Contact Information: Max Planck Institut für Radioastronomie, D-53902 Bad Münstereifel, Germany; phone: +49 2257-3010; fax: +49 2257-30169; Web site: http://www.mpifr-bonn.mpg.de/effberg.html

European Southern Observatory (ESO)

European organization, established in 1962, supported by Belgium, Denmark, France, Germany, Italy, the Netherlands, Sweden, and Switzerland. It has its headquarters near Munich, Germany, from where it acts as the technical and administrative center for observatories at La Silla, in the Chilean Andes, and Cerro Paranal, in the Atacama Desert in northern Chile. Telescopes at those two sites include:

Very Large Telescope (Cerro Paranal)

Four 8-m/26-ft reflectors mounted independently but capable of working in combination. This instrument was scheduled for full operation early in the twenty-first century.

La Silla

A 3.6-m/142-in telescope, commissioned in the mid-1970s. It was designed with interchangeable top units allowing it to be shifted between Cassegrain and Coude focus.

ESO/MPI Telescope (La Silla)

A 2.2-m/87-in Ritchey-Chretien telescope. It has been in operation since 1984 and is on loan to the ESO from the Max Planck Gesellschaft (the Max Planck Society for the Advancement of Science), an independent organization promoting scientific research, with headquarters in Munich, Germany.

Swedish-ESO Submillimeter Telescope (La Silla)

A 15-m/50-ft radio telescope built in 1987 on behalf of the Natural Science Research Council and the ESO.

Schmidt Telescope (La Silla)

A 1-m/39-in Cassegrain telescope, the first to be installed at the site. Since 1994, it has been dedicated to the Deep Near Infrared Southern Sky Survey (DENIS).

New Technology Telescope (La Silla)

An optical telescope with a thin, lightweight mirror, 3.5 m/138 in across. It is of rigid construction and is situated within a thermally controlled rotating building. It has active optical control to correct the defects and deformation of the telescope and mirror.

ESO Contact Information: Karl-Schwarzschild-Strasse 2, D-85748 Garching bei München, Germany; phone: +89 320-0600; fax: +89 320-2362; email: ips@eso.org; Web site: http://www.hq.eso.org/

Cerro Paranal Contact Information: ESO, Alonso de Cordova 3107, Casilla 19007, Santiago 19, Chile; phone: +56 55 281-291; fax: +56 55 242-680; email: ips@eso.org; Web site: http://www.eso.org/paranal/site/paranal.htm

La Silla Contact Information: ESO, Alonso de Cordova 3107; Casilla 19007, Santiago 19, chile; phone: +56 2 228-5006; fax: +56 2 228-5132; email: ips@eso.org; Web site: http://www.ls.eso.org/lasilla/generalinfo/html/aboutls.htm; ESO/MPI Telescope Web site: http://www.ls.eso.org/lasilla/Telescopes/2p2T/E2p2M/E2p2M.htm; Swedish-ESO Submillimeter Telescope Web site: http://www.ls.eso.org/lasilla/Telescopes/SEST/welcome.htm; Schmidt Telescope Web site: http://www.ls.eso.org/lasilla/Telescopes/denis/

html/1mtel.htm; New Technology Telescope Web site: http://
www.ls.eso.org/lasilla/Telescopes/NEWNTT/telescope/esontt.htm

European Space Agency (ESA)

Organization for the research and development of space applications
and technology. Member countries include Belgium, France, Germany,
Italy, the Netherlands, Britain, Denmark, Spain, Sweden, Switzerland,
Ireland, Austria, Norway, and Finland. The ESA space plan covers the fields
of science, Earth observation, telecommunications, space-segment technologies, ground infrastructures, and space transportation systems.

Contact Information: 8-10 Rue Mario Nikis, 75738 Paris Cedex,
France; phone: +33 1 5369 7654; fax: +33 1 5369 7560; Web site:
http://www.esa.int/

European Space Operations Center

European Space Agency (ESA) establishment, located in Darmstadt,
Germany. It is responsible for controlling and tracking ESA satellites
and spacecraft in orbit.

Contact Information: Robert-Bosch-Str. 5, Postbox 64276, 64293
Darmstadt, Germany; phone: +49 6151900; Web site: http://
www.esoc.esa.de/

European Space Research Institute

European Space Agency (ESA) establishment, located in Frascati, Italy.
It is responsible for ESA's Earth-observation satellites and for all of
its nonoperational data processing, information systems, and associated infrastructures, including the development of online catalogs and
directories.

Contact Information: Via Galileo Galilei, Casella Postale 64, 00044
Frascati, Italy; phone: +39 694 1801; fax: +39 694 180280; Web site:
http://www.esrin.esa.it/esrin/

European Space Technology Center

Technical center of the European Space Agency (ESA), located at
Noordwijk, Netherlands. The center is the largest of ESA's establish-
ments; its personnel are responsible for the technical preparation and
management of all ESA space projects and for providing technical sup-

port for ESA's satellite and manned space projects. The center is also the home of the agency's Space Science Department.

Contact Information: Keplerlaan 1, Postbus 299, 2200 AG Noordwijk, Netherlands; phone: +31 71 565 6040; fax: +31 71 565 6040; Web site: http://www.estec.esa.nl

Giant Metrewave Radio Telescope Facility

Radio astronomical research facility near Pune, India. Established by the National Center for Radio Astrophysics (NCRA), the site consists of thirty fully steerable parabolic dishes, each 45 m/148 ft in diameter, spread over distances of up to 25 km/15.5 mi.

Contact Information: Tata Institute of Fundamental Research, Poona University Campus, Post Bag 3, Ganeshkhind, Pune 411007, India; Web site: http://ncra.tifr.res.in/page/gmrt/gmrt.htm

Goddard Space Flight Center

Major field center of the National Aeronautics and Space Administration (NASA). Center personnel are responsible for the development and operation of unmanned scientific spacecraft. They also manage many of NASA's Earth-observation, astronomy, and space-physics missions and oversee operations at the Wallops Flight Facility, from which rockets carrying suborbital and low-Earth-orbit space probes are launched.

Contact Information: Greenbelt, MD 20771; phone: (301) 286-2000; email: goddard@compucom.com; Web site: http://pao.gsfc.nasa.gov/gsfc/welcome/welcome.htm

Green Bank Telescope Facility

U.S. telescope project located in Pocahontas County, West Virginia, under the auspices of the National Radio Astronomy Observatory (NRAO). The project was approved in 1990, and completion was expected in 1999. The 100-m/328-ft telescope was to be equipped with a laser ranging system, and the two thousand surface panels were to be individually mounted on motor-driven pistons to enable adjustment after construction.

Contact Information: P.O. Box 2, Green Bank, Pocahontas County, WV 24944; phone: (304) 456-2011; fax: (304) 456-2271; Web site: http://www.gb.nrao.edu/GBT/GBT.html

International Astronomical Union (IAU)

International organization for the promotion of the science of astronomy through international cooperation. The IAU, founded in 1919, has sixty-one country members. It is the sole internationally recognized authority for assigning designations to celestial bodies and their surface features. Its administrative center is the permanent Secretariat at the Institut d'Astrophysique in Paris, France.

Contact Information: IAU-UAI Secretariat, 98bis Bd. Arago, F-75014 Paris, France; phone: +33 1 4325 8358; fax: +33 1 4325 2616; Web site: http://www.iau.org/

Jet Propulsion Laboratory (JPL)

U.S. facility managed by the California Institute of Technology (Caltech) for the National Aeronautics and Space Administration (NASA). It is the primary U.S. center for robotic exploration of the solar system. JPL spacecraft have visited all known planets with the exception of Pluto.

Contact Information: 4800 Oak Grove Drive, Pasadena, CA 91109; phone: (818) 354-4321; Web site: http://www.jpl.nasa.gov

Johnson Space Center

National Aeronautics and Space Administration (NASA) facility established in 1961; home of mission control for crewed space flights. Named for former president Lyndon B. Johnson, it is NASA's primary center for the design, development, and testing of spacecraft. Center personnel are also responsible for the training of astronauts and the planning and implementation of crewed spaceflights.

Contact Information: 2101 Nasa Road 1, Houston, TX 77058-3696; phone: (281) 483-0123; email: majordomo@listserver.jsc.nasa.gov; Web site: http://www.jsc.nasa.gov

Kitt Peak National Observatory

Astronomical research facility, a division of the National Optical Astronomy Observatories (NOAO), located on Kitt Peak in the Quinlan Mountains of Arizona. The observatory operates numerous telescopes, including the 4-m/158-in Mayall reflector and the McMath-Pierce solar telescope. The latter actually comprises three telescopes, which may be used concurrently or independently: the main (1.61-m/63-in) and the east and west auxiliary (each 0.91-m/36-in) scopes. Also at Kitt Peak is a 12-m/39-ft millimeter-wavelength telescope operated by the National Radio Astronomy Observatory (NRAO). Upgraded from the 11-m/36-ft telescope in 1984, the facility enables a broad mix of studies, covering molecular clouds, galactic star formation, evolved stars, astrochemistry, and external galaxies, among other things.

Contact Information: NOAO, 950 N. Cherry Avenue, P.O. Box 26723, Tucson, AZ 85726-6732; phone: (520) 318-8726; fax: (520) 318-8360; email: kpno@noao.edu; Kitt Peak Observatory Web site: http://www.noao.edu/kpno/knpo.html; solar telescope Web site: http://www.nso.noao.edu/nsokp/mp.htm; 12-m telescope Web site: http://w.tuc.nrao.edu/telescope.htm

Kourou Space Center

Site of the *Ariane* space-launch facilities of the European Space Agency (ESA). Located in French Guiana, close to the equator, the site is ideal for launching geostationary satellites. ESA took over the existing European facilities at the Guiana Space Center in 1975 in order to build the *Ariane* launch complex.

Contact Information: Web site: http://www.esrin.esa.it/esa/ariane/facilities.html

Las Campanas Observatory

Site of a 2.5-m/100-in diameter telescope at Las Campanas, Chile, operated by the Carnegie Institution of Washington, D.C. (OCIW). A joint project by the OCIW, the University of Arizona, Harvard University, the University of Michigan, and the Massachusetts Institute of Technology (MIT) to add two 6.5-m/21.3-ft telescopes, known as the Magellan Project, began in 1992.

Contact Information: Las Campanas, Chile; phone: +56 51 224 680; email: johns@ociw.edu; Web site: http://www.ociw.edu/~johns/magellan.html

Marshall Space Flight Center

National Aeronautics and Space Administration (NASA) facility. It is NASA's primary center for the development and maintenance of space transportation and propulsion systems, including Saturn rockets and space-shuttle engines. The center also conducts microgravity research.

Contact Information: Web site: http://www.msfc.nasa.gov/

Mauna Kea Observatories

Astronomical observation site on the island of Hawaii, consisting of nine telescopes and the Hawaii Antenna of the Very Long Baseline Array. Under construction in the late 1990s were the Submillimeter Array and the Gemini Northern 8-Meter Telescope, which, with its twin 8-m/26-ft astronomical telescope at Cerro Pachon, Chile, will be able to cover the whole sky.

Contact Information: University of Hawaii, 2444 Dole Street, Honolulu, HI 96822; phone: (808) 956-8111; email: mkocn@ifa.hawaii.edu; Web site: http://www.ifa.hawaii.edu/mko/

Minor Planet Center

Nonprofit U.S. organization responsible for the efficient collection, checking, and circulation of astrometric observations and orbits for minor planets and comets. Information is circulated through the *Minor Planet Circulars* (monthly), and the *Minor Planet Electronic Circulars* (as necessary). The center, which operates at the Smithsonian Astrophysical Observatory, is principally funded by subscriptions to its various services.

Contact Information: Smithsonian Astrophysical Observatory, 60 Garden Street, Cambridge, MA 02138; phone: (617) 495-7280; fax: (617) 495-7231; email: iausubs@cfa.harvard.edu; Web site: http://cfa-www.harvard.edu/cfa/ps/mpc.html

MIT Haystack Observatory

Interdisciplinary research center for radio astronomy, geodesy, atmospheric sciences, and radar applications run by the Massachusetts Institute of Technology (MIT). The observatory is operated by the Northeast Radio Observatory Corporation (NEROC) under agreement with MIT. The principal instrument at the observatory is the 37-m/121-ft Haystack Radio Telescope enclosed in a radome.

Contact Information: Route 40, Westford, MA 01886; phone: (781) 981-5407; email: jes@wells.haystack.edu; Web site: http://sneezy.haystack.mit.edu/

Mount Graham International Observatory

International establishment located near Safford, Arizona. The site was chosen for its high altitude and low light pollution. It is the base for the 1.8-m/71-in Vatican Advanced Technology Telescope (VATT) and two others: the carbon-fiber 10-m/32.8-ft Heinrich Hertz Submillimeter Telescope (SMT), a joint project of the University of Arizona and the Max Planck Institute for Radio Astronomy, and the Large Binocular Telescope (LBT), with twin mirrors each measuring 8.4 m/133 in in diameter. The LBT, a collaboration among Arizona, Ohio, Italy, and Germany, is under construction; completion is expected in 2002.

Mount Graham Contact Information: Web site: http://medusa.as.arizona.edu/graham/graham.html; VATT Web site: http://clavius.as.arizona.edu/vo/vatt.html

SMT Contact Information: Steward Observatory, University of Arizona, Tucson, AZ 85721-0065; phone: (520) 621-5290; fax: (520) 621-5554; Web site: http://maisel.as.arizona.edu:8080/smt.html

LBT Contact Information: Steward Observatory, University of Arizona, Tucson, AZ 85721-0065; phone: (520) 626-5231; fax: (520) 621-9843; email: jhill@as.arizona.edu; Web site: http://medusa.as.arizona.edu/lbtwww/lbt.html

Mount Palomar Observatory

Owned and operated by the California Institute of Technology (Caltech), the observatory, located 70 km/43.5 mi north-northeast of San Diego, is used to support the scientific research programs of Caltech faculty and staff. The principal instruments are the 5-m/200-in Hale Telescope, the 1.22-m/48-in Oschin Telescope, the 0.5-m/18-in Schmidt Telescope, and the 1.5-m/60-in reflecting telescope.

Contact Information: 105–24 Caltech, 1201 E. California Blvd., Pasadena, CA 91125; phone: (626) 395-4169; fax: (626) 568-9523; observatory Web site: http://astro.caltech.edu/observatories/palomar/index.html

Mount Wilson Observatory

U.S. astronomical observation site, near Pasadena, California. Several ongoing observing projects are located here, including the 2.5-m/100-in Hooker Telescope. Built in 1917, it was named for John D. Hooker, a major donor to the project, and was the largest telescope of its kind for thirty years. In 1981, it was designated an International Historical Mechanical Engineering Landmark by the American Society of Mechanical Engineers.

Contact Information: Mount Wilson Institute, Hale Solar Laboratory, P.O. Box 60947, Pasadena, CA 91116; phone: (626) 793-3100; fax: (626) 793-4570; observatory Web site: http://www.mtwilson.edu/; telescope Web site: http://www.mtwilson.edu/Education/History/cal88/cal1188.html

Mullard Radio Astronomy Observatory

British astronomical facility, opened in 1972, operated under the auspices of the Cavendish Laboratory, Cambridge University, England. The observatory has several telescopes, including:

Cambridge Low-Frequency Synthesis Telescope (CLFST)

An east-west aperture telescope mounted on a 4.6-km/2.8-mi baseline.

Cambridge Optical Aperture-Synthesis Telescope (COAST)

A long-baseline interferometer array consisting of four telescopes. It was the first instrument of its kind to produce optical aperture-synthesis images (September 1995).

Ryle Telescope (RT)

Eight 12.8-m/42-ft Cassegrain aerials; four are mounted on a rail track, and the others are fixed at 1.2-km/0.7-mi intervals in a line 5 km/3 mi long. It has an east-west baseline.

Contact Information: Cavendish Laboratory, Madingley Road, Cambridge CB3 0HE, UK; phone: +44 1223 337294; fax: +44 1223 354599; observatory Web site: http://www.mrao.cam.ac.uk/; CLFST Web site: http://www.mrao.cam.ac.uk/telescopes/clfst/index.html; COAST Web site: http://www.mrao.cam.ac.uk/telescopes/coast/coast.intro.html; RT Web site: http://www.mrao.cam.ac.uk/telescopes/ryle/index.html

National Aeronautics and Space Administration (NASA)

U.S. agency for scientific research and aerospace exploration. NASA was founded in 1958, partly due to pressures of national defense during the cold war with the Soviet Union, and expanded rapidly, beginning in the 1960s with the Mercury, Gemini, and Apollo projects.

Contact Information: 300 E Street, SW, Washington, DC 20546; phone: (202) 358-0162; email: help@sti.nasa.gov/; Web site: http://www.nasa.gov/

National Radio Astronomy Observatory

Astronomical facility of the National Science Foundation located in Charlottesville, Virginia.

Contact Information: 520 Edgemont Road, Charlottesville, VA 22903-2475; phone: (804) 296-0211; fax: (804) 296-0278; Web site: http://www.cv.nrao.edu/

National Space Science Data Center

Division of the Goddard Space Flight Center. Center personnel provide access to a wide variety of astrophysics, space physics, lunar, and planetary data for the National Aeronautics and Space Administration's (NASA) spaceflight missions.

Contact Information: NASA Goddard Space Flight Center, Code 633, Greenbelt, MD 20771; phone: (301) 286-9789; email: nate.james@gsfc.nasa.gov; Web site: http://nssdc.gsfc.nasa.gov/nssdc/nssdc_home.htm

Parkes Observatory

Australian observatory, part of the Australia Telescope National Facility (ATNF), a division of the Commonwealth Scientific and Industrial Research Organization (CSIRO). The site contains a 64-m/210-ft radio telescope.

Contact Information: P.O. Box 276, Parkes, New South Wales 2870, Australia; phone: +61 2 6861-1700; fax: +61 2 6861-1730; email: parkes@atnf.csiro.au; Web site: http://www.parkes.atnf.csiro.au/library/MANUALS/OBG/OBG_2.htm

Paul Wild Observatory

Australian observatory, part of the Australia Telescope National Facility (ATNF), operated by the Commonwealth Scientific and Industrial Research Organization (CSIRO). The site contains the Australia Telescope Compact Array, consisting of a total of six antennae at Narrabri, New South Wales.

Contact Information: CSIRO, Australia Telescope National Facility, Paul Wild Observatory, Locked Bag 194, Narrabri, New South Wales 2390, Australia; phone: +61 2 6790-4000; fax: +61 2 6790-4090; email: narrabri@atnf.csiro.au; Web site: http://www.narrabri.atnf.csiro.au

Royal Astronomical Society (RAS)

British society for the promotion of astronomy and geophysics. Its main function is to publish the results of astronomical and geophysical research, and its library at Burlington House, London, contains extensive historical and research material in these fields. The society, founded in 1820, is affiliated with the European Astronomical Society and the European Geophysical Society. The *Royal Astronomical Society Monthly Notices,* published for the RAS by Blackwell Science Ltd., presents the results of original research in positional, dynamical, and radio astronomy, astrophysics, cosmology, space research, and the design of astronomical instruments.

Contact Information: Burlington House, Piccadilly, London W1V 0NL, UK; phone: +44 171 734 3307; fax: +44 171 494 0166; email: info@ras.org.uk; Web site: http://www.ras.org.uk/ras/

Monthly Notices Editorial Office: P.O. Box 88, Oxford, OX2 0NE, UK; phone: 44 1865 206126; fax: 44 1865 206219; email: journals.cs@blacksci.co.uk; Web site: http://www.blacksci.co.uk/products/journals/mnras.htm

Russian Space Agency

Federal body established in 1992 by decree of the president of the Russian Federation. Its primary purpose is to define state policy for space research and exploration for peaceful purposes. It aims to provide effective solutions to social and economic tasks and to promote Russia's interests in the international space arena.

Contact Information: 42 Shchepkin Street, Moscow 129857, Russia; email: www@rka.ru; Web site: http://www.rka.ru/

Smithsonian Astrophysical Observatory (SAO)

U.S. organization for research in astronomy, astrophysics, Earth and space sciences, and science education. SAO contains seven different scientific divisions, as well as many other archives and services. Publications include the *Preprint Series* and a quarterly *Almanac*. The SAO is joined with the Harvard College Observatory to form the Harvard-Smithsonian Center for Astrophysics (CfA).

Contact Information: Harvard-Smithsonian Center for Astrophysics, 60 Garden St., Cambridge, MA 02138; email: pubaffairs@cfa.harvard.edu; Web site: http://cfa-www.harvard.edu/saohome.htm

Subaru Telescope Facility

Optical-infrared 8.3-m/27.2-ft telescope at the summit of Mauna Kea, Hawaii. Operated by the National Astronomical Observatory (NAO) under the Ministry of Education of Japan, the site was due to be fully operational in the year 2000.

Contact Information: 650 North A'ohoku Place, Hilo, HI 96720; phone: (808) 934-7788; fax: (808) 934-5099; Web site: http://chain.mkt.nao.ac.jp/subaru.htm

University of California/Lick Observatory

Observatory of the University of California and Mount Hamilton, California. Its primary research instruments are the 3.04-m/120-in Shane reflector and a 91-cm/36-in refractor, the second-largest refractor in the world.

Contact Information: University of California, Santa Cruz, CA 95064; phone: (831) 459-2513; fax: (831) 426-3115; email: webeditor@ucolick.org; Web site: http://www.ucolick.org

Very Large Array (VLA)

U.S. astronomical radio observatory west of Socorro, New Mexico. It is operated by the National Radio Astronomy Observatory (NRAO). Construction began in 1973, and the array was completed in 1980. The VLA comprises twenty-seven antennae, each one measuring 25 m/82 ft in diameter; combined, they give the resolution of an antenna 27 km/17 mi across.

Contact Information: P.O. Box 0, Socorro, NM 87801; phone: (505) 835-000; fax: (505) 835-027; email: dfinley@nrao.edu; Web site: http://www.nrao.edu/vla/html/VLAhome.shtml

Very Long Baseline Array (VLBA)

Series of ten radio telescopes situated at stations across the continental United States, and on Mauna Kea, Hawaii, and St. Croix in the U.S.

Virgin Islands. The array is operated by the National Radio Astronomy Observatory and controlled by computer from the Array Operations Center in Socorro, New Mexico. Construction of the VLBA began in 1986, and the last station, on Mauna Kea, was completed in 1993.

Contact Information: P.O. Box 0, Socorro, NM 87801; phone: (505) 835-000; fax: (505) 835-027; Web site: http://www.nrao.edu/intro/NM_astro.html

WIYN Observatory

Site of the 3.5-m/138-in WIYN Telescope at Kitt Peak, Arizona. The observatory is owned and operated by the WIYN Consortium, consisting of the University of Wisconsin, Indiana University, Yale University, and the National Optical Astronomy Observatories (NOAO). The telescope is the second-largest on the site and employs many technological innovations.

Contact Information: NOAO, 950 N. Cherry Avenue, P.O. Box 26723, Tucson, AZ 85726-6732; phone: (520) 318-8726; fax: (520) 318-8360; email: kpno@noao.edu; Web site: http://www.noao.edu/wiyn/wiynis.htm

W. M. Keck Observatory

Astronomical observation site on Mauna Kea, Hawaii. Operated by California Institute of Technology, the University of California, and the National Aeronautics and Space Administration (NASA), the site contains the twin Keck Telescopes, each containing a 10-m/33-ft primary mirror.

Contact Information: email: www@keck.hawaii.edu; Web site: http://www2.keck.hawaii.edu:3636/

5

Selected Works for Further Reading

Allen, Richard. *Star Names: Their Lore and Meaning.* New York: Dover, 1963. 563 pages. ISBN: 0486210790.

A standard and comprehensive survey of the meaning of star names in English, Greek, Arabic, and other languages.

Ashbrook, John. *The Astronomical Scrapbook: Skywatchers, Pioneers, and Seekers in Astronomy.* Cambridge, UK; New York: Cambridge University Press, 1984. 468 pages. ISBN: 0521300452.

A fascinating collection of ninety-one short articles, mainly gleaned from the more curious byways of the history of astronomy, that originally appeared in the popular magazine *Sky and Telescope.*

Bahcall, John N., and Jeremiah P. Ostriker, eds. *Unsolved Problems in Astrophysics.* Princeton, NJ: Princeton University Press, 1997. 377 pages. ISBN: 0691016070.

Different authors tackle various challenging questions in the field, with the stress on the questions still to be answered. May be a little daunting for the general reader.

Barrow, John, and Frank Tipler. *The Anthropic Cosmological Principle.* Oxford, UK; New York: Oxford University Press, 1986. 706 pages. ISBN: 0198519494.

An exhaustive overview of the cosmos and humankind's place in it, daunting in parts but quite readable if you skip the technical stuff.

Brandt, John C. *Comets: Readings from* Scientific American. San Francisco: W. H. Freeman, 1981. 92 pages. ISBN: 0716713195.

A collection of authoritative articles covering such topics as the tails, spin, and nature of comets. A simpler account can be found in *Guide to Comets* (1977) by Patrick Moore and Peter Cattermole.

Burke, Bernard F., and Francis Graham-Smith. *An Introduction to Radio Astronomy.* New York: Cambridge University Press, 1997. 297 pages. ISBN: 0521554543.

An excellent textbook covering both the technical details of radio telescopes and their view of the universe.

Cattermole, Peter, and Patrick Moore. *Atlas of Venus.* New York: Cambridge University Press, 1997. 143 pages. ISBN: 0521496527.

Readable, detailed, and extremely well illustrated.

Chown, Marcus. *Afterglow of Creation: From the Fireball to the Discovery of Cosmic Ripples.* London: Arrow, 1993. 173 pages. ISBN: 0099280515.

The best book about the cosmic background radiation that fills all of space and is a remnant of the Big Bang itself.

Clark, Stuart. *Redshift.* London: University of Hertfordshire Press, 1997. ISBN: 0900458798.

Good, clear undergraduate explanation.

Clark, Stuart. *Universe in Focus: The Story of the Hubble Telescope.* London: Cassell, 1997. 128 pages. ISBN: 0304349453.

Contains 150 high-quality reproductions of Hubble images, accompanied by details of Hubble's instruments and a wealth of astronomical detail.

Doran, Jamie, and Piers Bizony. *Starman.* London: Bloomsbury, 1998. 248 pages. ISBN: 0747536880.

A lively account of the life of Soviet cosmonaut Yuri Gagarin, who, on April 12, 1961, became the first person to enter space, including much information long suppressed.

Freeman, W. H., ed. *Solar System: Readings from* Scientific American. San Francisco: W. H. Freeman, 1975. 145 pages. ISBN: 0716705516.

A useful and readable collection of articles on the solar system, including individual papers on the planets, their origin, and interplanetary fields.

Gamow, George. *The Creation of the Universe.* New York: Viking Press, 1952. 147 pages.

Fascinating, if slightly dated, "horse's mouth" account from one of the pioneers of the Big Bang theory.

Gingerich, Owen, ed. *Frontiers in Astronomy: Readings from* Scientific American. San Francisco: W. H. Freeman, 1970. 370 pages. ISBN: 0716709481.

Gingerich, Owen, and Paul W. Hodge, eds. *The Universe of Galaxies.* New York: W. H. Freeman, 1984. 113 pages. ISBN: 0716716755.

The two works are valuable collections of articles from *Scientific American* on such topics as quasars, dark matter, the Milky Way, the red shift, and exploding galaxies.

Gleiser, Marcelo. *The Dancing Universe: From Creation Myths to the Big Bang.* New York: Dutton, 1997. 338 pages. ISBN: 0525941126.

Combining popular science with humor, the author takes a wide view of the origins of the universe.

Gribbin, John. *In Search of the Big Bang: Quantum Physics and Cosmology.* Toronto; New York: Bantam Books, 1986. 413 pages. ISBN: 0553342584.

Historical account of the development of cosmology in the twentieth century, from the expanding universe to the theory of inflation.

Gribbin, John, and Martin Rees. *The Stuff of the Universe: Cosmic Coincidences: Dark Matter, Mankind, and Anthropic Cosmology.* London: Black Swan, 1991. 302 pages. ISBN: 055299443X.

British Astronomer Royal Sir Martin Rees joins forces with science writer John Gribbin to describe the dark matter that makes up 99 percent of the universe.

Grinspoon, David. *Venus Revealed: A New Look Below the Clouds of Our Mysterious Twin Planet.* Reading, MA: Addison-Wesley, 1997. 355 pages. ISBN: 0201406551.

Nontechnical, clear, and thorough account.

Grosser, Morton. *The Discovery of Neptune*. New York: Dover, 1979. 172 pages. ISBN: 0486237265.

A fascinating account of the discovery in 1846 of the existence of a previously unknown planet.

Hale, Alan. *Everybody's Comet: A Layman's Guide to Comet Hale-Bopp*. Silver City, NM: High-Lonesome Books, 1996. ISBN: 0944383386.

Everything you ever wanted to know about Hale-Bopp by its codiscoverer. It also provides a general introduction to comets.

Harford, James. *Korolev: How One Man Masterminded the Soviet Drive to Beat America to the Moon*. New York: Wiley, 1997. 392 pages. ISBN: 0471148539.

An account of his Sergei Korolev's life and works that also provides insight into what it was like to work in the USSR at the height of its power.

Harrington, Philip. *Eclipse: The What, Where, When, Why, and How Guide to Watching Solar and Lunar Eclipses*. New York: Wiley, 1997. 280 pages. ISBN: 0471127957.

Informative and enthusiastic.

Henbest, Nigel. *Observing the Universe*. Oxford: Blackwell & New Scientist. 288 pages. ISBN: 0855207272.

A collection of short articles from the British periodical *New Scientist* dealing with research into X-ray astronomy, ultraviolet astronomy, the gamma-ray sky, cosmic rays, the infrared sky, and optical and radio astronomy.

Henbest, Nigel, and Heather Couper. *The Guide to the Galaxy*. Cambridge, UK; New York: Cambridge University Press, 1994. 265 pages. ISBN: 0521306221.

A lavishly illustrated, popular, but detailed work on the Milky Way. It has chapters on the discovery of the Milky Way, its geography, its center, and the Perseus, Orion, and Sagittarius arms.

Hogan, Craig. *The Little Book of the Big Bang: A Cosmic Primer*. New York: Copernicus, 1998. 181 pages. ISBN: 0387983856.

Concisely written and useful look at the main questions of modern cosmology, with tables and diagrams.

Hoskins, M., and O. Pedersen, eds. *Gregorian Reform of the Calendar.* Vatican City: Pontificia Academia Scientarium,1983. 321 pages.

A full and fascinating account of the historical and astronomical problems of introducing the gregorian calendar into Europe.

King, Henry. *History of the Telescope.* New York: Dover, 1979. 456 pages. ISBN: 0486238938.

The standard history of the telescope from the earliest times to the 1950s.

Lancaster-Brown, Peter. *Megaliths, Myths, and Men: An Introduction to Astro-Archaeology.* New York: Taplinger, 1976. 324 pages. ISBN: 0800851870.

A critical and expert survey of the supposed astronomical background of Stonehenge, the Pyramids, and several other ancient remains. Lancaster-Brown disposes of some of the more outlandish claims commonly made about ancient astronomy.

Lewis, David. *The Voyaging Stars: Secrets of the Pacific Island Navigators.* New York: W. W. Norton, 1978. 208 pages. ISBN: 0393032264.

An account of how, throughout Oceania, islanders in traditional canoes navigate from one small island to an equally small, distant island.

Lewis, H. A. *Times Atlas of the Moon.* London: Times Newspapers, 1969. 110 pages. ISBN: 0723000069.

A detailed map of the moon. Less detailed but with more general information about the moon—its origin, orbit, and the like—is Patrick Moore's *The Moon* (1985).

Lovell, Bernard. *The Story of Jodrell Bank.* New York: Harper & Row, 1968. 265 pages.

The story of Lovell's attempts to design, finance, and build a 76-m/250-ft steerable parabolic radio telescope. Against all the odds, it was opened in 1957. More detail about the various telescopes built by Lovell can be found in *The Jodrell Bank Telescopes* (1985).

Mather, John C. *The Very First Light: The True Inside Story of the Scientific Journey Back to the Dawn of the Universe*. New York: Basic Books, 1996. 328 pages. ISBN: 0465015751.

> The story of the COBE Project—the U.S. *Cosmic Background Explorer* (COBE) satellite launched in 1989 to study microwave background radiation—and a wide-ranging foray into cosmology that is recommended for all levels.

Mitton, Jacqueline. *Concise Dictionary of Astronomy*. Oxford, UK; New York: Oxford University Press, 1991. 423 pages. ISBN: 0198539673.

> A clear, well-organized "A to Z" that includes a tremendous amount of information and is easy to digest.

Moore, Patrick. *Exploring the Night Sky with Binoculars*. Cambridge, UK; New York: Cambridge University Press, 1986. 203 pages. ISBN: 0521307562.

> A book for beginners. After offering advice on choosing binoculars, Moore writes about the planets, comets, and the moon. Half of the book is devoted to charts of the constellations.

Moore, Patrick. *The Guinness Book of Astronomy*. London: Guinness, 1992. 288 pages. ISBN: 0851129404.

> A familiar and informative guide to the solar system and the stars, with a ninety-page star catalog and brief sections on the history of astronomy.

Muirden, James. *Astronomy with a Small Telescope*. London: Hamlyn Octopus, 1985. ISBN: 0540011959.

> Introductory chapters describe the various telescopes and their mounting and are followed by advice on how to observe the sun, the moon, the planets, meteorites, comets, constellations, and galaxies. There is also a chapter on photography.

Muller, Richard. *Nemesis*. New York: Weidenfeld & Nicolson, 1988. 193 pages. ISBN: 1555841732.

> The exciting story of the search by astrophysicists in the 1980s for the "killer star" Nemesis. Nemesis is thought to be orbiting the sun and—as it approaches the earth every twenty-six million years—causing such catastrophes as the extinction of the dinosaurs.

Murdin, Paul. *End in Fire: The Supernova in the Large Magellanic Cloud.* Cambridge, UK; New York: Cambridge University Press, 1990. 251 pages. ISBN: 0521374952.

> Murdin describes the discovery in 1990 of SN1987A, the first supernova to be visible to the unaided eye since 1604, traces the development of the supernova, and shows its connection to theory.

Murdin, Paul, and Lesley Murdin. *Supernovae.* Cambridge, UK; New York: Cambridge University Press, 1985. 185 pages. ISBN: 052130038X.

> The Murdins begin their popular work with descriptions of the supernovae of 1066, 1572, and 1604 and go on to show how supernovae are related to pulsars, black holes, neutron stars, and the creation of the elements.

Nahin, Paul. *Time Machines: Time Travel in Physics, Metaphysics, and Science Fiction.* New York: American Institute of Physics, 1993. 408 pages. ISBN: 0883189356.

> Goes beyond the black hole, mixing science fiction and science fact to discuss the extraordinary implication of Albert Einstein's general theory of relativity: that time travel may not be impossible.

North, John. *The Norton History of Astronomy and Cosmology.* New York: W. W. Norton, 1994. ISBN: 0393036561.

> An authoritative and readable history (part of the Fontana History of Science series), this book covers astronomy from prehistory to Stephen Hawking, the British theoretical physicist whose books and PBS series have introduced millions to the mysteries of the universe.

Norton, Arthur Philip. *Norton's 2000: Star Atlas and Reference Handbook.* New York: Wiley, 1989. 179 pages. ISBN: 0470214600.

> *Norton's Atlas* first appeared in 1910; this latest edition (eighteenth) is calculated for the year 2000. A similar approach can be found in the *Cambridge Star Atlas 2000* (1991) by Will Trion.

Overbye, Dennis. *Lonely Hearts of the Cosmos.* New York: HarperCollins, 1991. 438 pages. ISBN: 0060159642.

> The story of the quest for the secret of the origin of the universe, told in terms of the personalities involved. Overbye gives a real flavor of how cosmologists work and think.

Padmanabhan, T. *After the First Three Minutes: The Story of Our Universe*. Cambridge, UK; New York: Cambridge University Press, 1998. 215 pages. ISBN: 0521620392.

> An enjoyable account of galaxy formation and evolution but with all of the physics, too.

Phillips, Kenneth. *Guide to the Sun*. Cambridge, UK; New York: Cambridge University Press, 1992. 386 pages. ISBN: 052139483X.

> Though somewhat technical in parts, this book provides a readable account of the chromosphere, photosphere, corona, and interior of the sun and how best it can be observed.

Plant, Malcolm. *Dictionary of Space*. Harlow, Essex, UK: Longman, 1986. 270 pages. ISBN: 0582892961.

> An essential aid for those who cannot remember whether *Apollo 7* went to Mars or the moon or which rocket first landed on Venus.

Ronan, C. A. *The Natural History of the Universe: From the Big Bang to the End of Time*. New York: MacMillan, 1991. 212 pages. ISBN: 0026045117.

> A well-written, beautifully illustrated general survey of modern astronomy; it has been widely acclaimed.

Sheehan, William. *The Planet Mars: A History of Observation and Discovery*. Tucson: University of Arizona Press, 1996. ISBN: 0816516405.

> Readable, nontechnical history of Martian discoveries.

Smith, Robert. *The Expanding Universe: Astronomy's "Great Debate," 1900–1931*. Cambridge, UK; New York: Cambridge University Press, 1982. 220 pages. ISBN: 0521232120.

> A scholarly but readable account of the theoretical arguments and observational evidence that enabled astronomers such as the American Edwin Hubble to conclude that the universe is expanding. The same period and argument are covered in the more popular and lavishly illustrated *Man Discovers the Galaxies* (New York: Columbia University Press, 1984; ISBN: 0231058268) by R. Berendzen, R. Hart, and Daniel Seeley.

Stern, Alan, and Jacqueline Mitton. *Pluto and Charon: Ice Worlds on the Ragged Edge of the Solar System*. New York: Wiley, 1998. 216 pages. ISBN: 0471152978.

An excellent roundup of all that is known of Pluto and its moon.

Stott, Carol, ed. *Images of the Universe*. Cambridge, UK; New York: Cambridge University Press, 1991. 237 pages. ISBN: 0521391784.

A book for those who observe the heavens; each chapter is written by an expert in his particular field. Most of the authors are closely linked with the British Astronomical Association.

Sullivan, Walter. *We Are Not Alone: The Search for Intelligent Life on Other Worlds*. New York: McGraw-Hill, 1966. 325 pages. ISBN: 0070623244.

A sensible and interesting discussion of whether life exists on worlds outside our solar system.

Tayler, Roger. *The Hidden Universe*. New York: E. Howard, 1991. 213 pages. ISBN: 0133887111.

A very clear account of some of the problems of modern cosmology.

Tucker, Wallace, and Riccardo Giacconi. *The X-Ray Universe*. Cambridge, MA: Harvard University Press, 1985. 201 pages. ISBN: 0674962850.

X-ray stars were first observed in 1962. The authors describe the discovery and the resulting research that led to the launch of two X-ray telescopes, *Uhuru* in 1970 and *Einstein* in 1978, and the results gathered.

Wali, Kameshwar, ed. *S. Chandrasekhar*. London: Imperial College Press, 1998. 236 pages. ISBN: 1860940382.

Collection of thirty-five essays of reminiscence and tribute written by friends and colleagues of the Indian-born U.S. astrophysicist Subrahmanyan Chandrasekhar.

Whipple, Fred. *Orbiting the Sun: Planets and Satellites of the Solar System*. Cambridge, MA: Harvard University Press, 1981. 338 pages. ISBN: 0674641256.

A comprehensive account of the solar system, incorporating satellite data and photographs from the U.S. *Viking* missions. Patrick Moore's *New Guide to the Planets* (1993) covers much the same ground at a less technical level.

Will, Clifford. *Was Einstein Right? Putting General Relativity to the Test.* New York: Basic Books, 1986. 274 pages. ISBN: 0465090885.

The answer, of course, is yes. This book is the most accessible guide to Einstein's theories for the nonscientist.

Williams, J. E. D. *From Sails to Satellites: The Origins and Development of Navigational Science.* Oxford, UK; New York: Oxford University Press, 1992. 310 pages. ISBN: 0198563876.

Williams tells the story of navigational science from the first attempts as sea navigation to recent times, including the use of radio and radar in navigation.

Worvill, Roy. *Telescope Making for Beginners.* New York: Orbiting Book Service, 1974. 79 pages. ISBN: 0914326007.

For those prepared to make their own telescope, the author offers a simple guide.

Zeilik, M., and J. Gaustad. *Astronomy: The Cosmic Perspective.* New York: Wiley, 1990. 822 pages. ISBN: 047150016X.

A major single-volume textbook dealing with the evolution of the stars and the galaxies. Though technical in parts, much of the text remains within the competence of the general reader.

6　World Wide Web Sites

About Goddard Space Flight Center
http://pao.gsfc.nasa.gov/gsfc/welcome/history/history.htm

> Primarily composed of a biography and an achievement history of the American rocket pioneer Robert Hutchings Goddard, the Web site also describes the Goddard Space Flight Center, Greenbelt, Maryland, and details the type of mission it is responsible for undertaking. The site, part of NASA's (National Aeronautics and Space Administration) huge Internet network of resources, contains many links to related topics.

Advanced Communication Technology Satellite (ACTS)
http://kronos.lerc.nasa.gov/acts/acts.html

> This site provides project information on the ACTS project, run by NASA's Lewis Research Center, Cleveland, Ohio. It includes detailed descriptions of ACTS spacecraft, operations, and experiments and the latest news and related events.

AJ's Cosmic Thing
http://www.mindentimes.on.ca/CosmicThing/Main.html

> Set any time, date, longitude, latitude, and position in the setting window, and this impressive plotter applet will display stationary or moving full-sky images of objects in the sky. Click on plotted objects and get information about them. There is a technical page for those wishing additional information.

Apollo 11
http://www.nasa.gov/hqpao/apollo_11.html

> This NASA page relives the excitement of the 1969 *Apollo 11* mission to the moon, with recollections from the participating astronauts,

images, audio clips, access to key White House documents, and a bibliography.

Ask an Astronaut
http://www.nss.org/askastro/

This is a multimedia archive of the 1972 *Apollo 17* mission to the moon, with video and audio files, mission details, and astronaut biographies. Sponsored by the National Space Society, it includes numerous images, a list of questions and answers, and links to related sites.

Ask the Space Scientist
http://image.gsfc.nasa.gov/poetry/ask/askmag.html

This well-organized site provides answers to questions about astronomy and space. If you can't find an answer to your question in the site's question archives, you can e-mail your question to site staff, who will endeavor to send a reply within three days. There are useful links to question archives, general files, and other space-related Web sites.

Asteroid and Comet Impact Hazards
http://impact.arc.nasa.gov/index.html

This NASA site offers an overview and the latest news on asteroid- and comet-impact hazards from the agency's Ames Space Science Division, along with the last Spaceguard Survey Report and a list of future near-Earth objects (NEOs).

Asteroid Introduction
http://www.hawastsoc.org/solar/eng/asteroid.htm

What is the difference between an asteroid and a meteorite? Find out at this site, which contains a table of statistics about asteroids, a chronology of asteroid exploration, and images of selected asteroids.

Brief History of Cosmology
http://www-history.mcs.st-and.ac.uk/~history/HistTopics/
Cosmology.html

Based at St. Andrews University, Scotland, this site chronicles the history of cosmology from the time of the Babylonians to the Hubble Space Telescope. It includes links to the biographies of the key historical figures responsible for the advancement of the subject and a brief list of references for further reading.

Cassini Mission to Saturn
http://miranda.colorado.edu/cassini/

Well-presented information on the purposes of the *Cassini* space probe, launched in 1997 by NASA and the European Space Agency to the planet Saturn. The complexities of designing the probe and its instruments and navigating it to the distant planet are explained in easy-to-understand language with the aid of good graphics. There are also interesting details of what is already known about Saturn's moons.

Challenger Remembered: The Shuttle Challenger's Final Voyage
http://uttm.com/space/challenger.html

The information at this site was prepared by the Cape Canaveral, Florida, bureau chief for United Press International, who witnessed and reported on the January 28, 1986, *Challenger* disaster, which killed all seven members of the U.S. space-shuttle crew, and the ensuing investigation. Four sections provide detailed information on the events leading up to the *Challenger* launch and the immediate aftermath, including a time line with transcriptions of the communications between *Challenger* and mission control, information regarding the fate of the crew (included to dispel myths and misinformation), and the details of the report of the Rogers Commission, which investigated the tragedy.

Chandrasekhar, Subrahmanyan
http://www-groups.dcs.st-and.ac.uk/history/Mathematicians/
Chandrasekhar.html

Part of an archive at the University of St. Andrews, Scotland, containing the biographies of the world's greatest mathematicians, this site is devoted to the life and contributions of Indian-born U.S. astronomer Subrahmanyan Chandrasekhar. In addition to biographical information, you will find a list of references about Chandrasekhar and links to other essays in the archive that reference him. The text of the essay at this site includes hypertext links to essays of mathematicians and thinkers who influenced Chandrasekhar. You will also find an image of him, which you can click on to enlarge, and a map of his birthplace, Lahore (now in Pakistan).

Clyde Tombaugh Profile
http://www.achievement.org/autodoc/page/tom0pro-1

At this site, you will find a description of the life and work of the discoverer of Pluto, U.S. astronomer Clyde Tombaugh. In addition to a profile and biographical information, the Web site also contains a lengthy interview with Tombaugh from 1991, accompanied by a large number of photographs, video sequences, and audio clips.

Comet Hale-Bopp Home Page
http://www.jpl.nasa.gov/comet/

Contains nearly five thousand images of the comet, which was visible with the naked eye when it came within 190 million km/118 million mi of Earth in 1997, information on its discovery and position, and details of current research findings.

Comet Introduction
http://www.hawastsoc.org/solar/eng/comet.htm

Learn how to make a comet nucleus at this site, which also contains a diagram and explanation of their orbits, a chronology of comet exploration, and information and photographs of selected comets.

Comet Shoemaker-Levy Home Page
http://www.jpl.nasa.gov/sl9/

This site provides a description of the comet's collision with Jupiter in 1994, the first collision of two solar-system bodies ever to be observed. Background information, the latest theories about the effects of the collision, and even some animations of Jupiter and the impact are included.

Constellations and Their Stars
http://www.vol.it/mirror/constellations/

This site offers notes on the constellations (listed alphabetically, by month, and by popularity), plus lists of the twenty-five brightest and the thirty-two nearest stars and photographs of the Milky Way.

Earth and Moon Viewer
http://www.fourmilab.ch/earthview/vplanet.html

View a map of Earth showing the day and night regions at the moment you visit this site, or view Earth from the sun, the moon, or any number of other locations. You can also take a look at the moon from Earth or the sun or from above various formations on the lunar surface.

Europa
http://www.hawastsoc.org/solar/eng/europa.htm

This site provides the latest comprehensive information on Jupiter's moon. Included are a full discussion of evidence that Europa may have running water on its surface and a number of false and natural color images of Europa.

European Southern Observatory (ESO)
http://www.eso.org/

Reports from the intergovernmental European observatory. Contents include descriptions of ESO observing facilities and the structure of ESO, press releases, scientific reports, lists of ESO publications, and images of space.

Face of Venus Home Page
http://www.eps.mcgill.ca/~bud/craters/FaceOfVenus.html

This Web site gives an overview and description of the surface of the planet Venus. It includes interactive databases of coronas and craters, as well as many images.

FAQ on Telescope Buying and Usage
http://www-personal.umich.edu/~dnash/astrodir/saafaq/faq.html

You will find here an extensive "plain English" guide to buying a telescope. For anybody interested in astronomy and contemplating buying a telescope or setting up an observatory, this is an indispensable source of noncommercial advice. Included are a full explanation of jargon used in astronomy and guides to help the novice amateur astronomer.

Galileo Project Information
http://nssdc.gsfc.nasa.gov/planetary/galileo.html

This site dedicated to the U.S. Galileo Project to explore Jupiter and its moons and the opportunities it has offered scientists and astronomers to enhance our understanding of the universe. The site includes information about the mission's objectives and scientific results, images, and links to other relevant sites on the Web.

Gong Show: Ringing Truths about the Sun
http://www.sciam.com/explorations/072896explorations.html

Part of a larger site maintained by the publication *Scientific American,* this page reports on the fledgling scientific field of helioseismology, the study of the sun's interior. Find out what scientists are learning about the processes taking place deep within the sun. The text includes hypertext links to further information, and there is also a list of related links at the bottom of the page.

History of Mount Wilson Observatory
http://www.mtwilson.edu/History/

This Web site features many articles and images relating to the history of the observatory, which is located near Pasadena, California. It includes images of visits to its 2.5-m/100-in telescope by such luminaries as German-born U.S. physicist Albert Einstein and U.S. astronomer George Ellery Hale.

Inquirer's Guide to the Universe
http://sln.fi.edu/planets/planets.html

This Web site is designed for teachers and school students, with pages on "space science fact" (the universe as humans know it today) and "space science fiction" (the universe as humans imagine it might be). Features include planetary fact sheets, information about planets outside our solar system, virtual trips to black holes and neutron stars, space quotes, and a course in spaceship design.

Jodrell Bank Home Page
http://www.jb.man.ac.uk/index.html

This comprehensive site describes the Nuffield Radio Astronomy Laboratories at Jodrell Bank, Cheshire, England. The radio telescope has been used for astronomy for many years; this site describes some of the important discoveries it has led to and also hints at future developments.

Jupiter
http://www.hawastsoc.org/solar/eng/jupiter.htm

You will find here full details of the planet and its moons, including a chronology of exploration, various views of the planet and its moons, and links to other planets.

Kennedy Space Center
http://www.ksc.nasa.gov/

This is NASA's well-presented guide to the history and current operations of the U.S. gateway to the universe. There is an enormous quantity of textual and multimedia information of interest to both the general reader and those who are technically inclined.

Kitt Peak National Observatory
http://www.noao.edu/kpno/kpno.html

Comprehensive information on the range of research carried out at Kitt Peak, near Tucson, Arizona. In addition to scientific information of interest mainly to professional astronomers, there are details of current weather conditions at Kitt Peak and information for visitors. The site also provides access to visible and infrared images from satellites in geostationary orbits.

Learning Center for Young Astronomers
http://heasarc.gsfc.nasa.gov/docs/StarChild/StarChild.html

This Web site introduces young astronomers to the universe. The presentation covers a wide range of issues, with discussions of quasars, comets, meteoroids, the Milky Way, black holes, the Hubble Space Telescope, space wardrobes, and space probes. The tutorial is offered in two levels—one basic and one more advanced—and also includes online activities, visual material, and an illuminating glossary of space terms.

Mars
http://www.hawastsoc.org/solar/eng/mars.htm

Presented at this site is a detailed description of the planet Mars, commonly referred to as the Red Planet. It includes statistics and information about the surface, volcanoes, satellites, and clouds of the planet, supported by a good selection of images.

Mauna Kea Observatories
http://www.ifa.hawaii.edu/mko/mko.html

This is the official site of the renowned international observatories. A clickable photo on the home page accesses information about the functions and findings of the group of telescopes atop Hawaii's highest peak. In addition to scientific data, the site also contains information for visitors to the facilities.

Mercury
http://www.hawastsoc.org/solar/eng/mercury.htm

This detailed description of the planet Mercury includes statistics and information about the planet, along with a chronology of its exploration supported by a good selection of images.

Meteors, Meteorites, and Impacts
http://www.seds.org/billa/tnp/meteorites.html

This site offers an informative collection of facts about meteorites: their classification, how they are formed, and what happens when one hits Earth. It includes images of a selection of meteorites.

Moon
http://www.hawastsoc.org/solar/eng/moon.htm

A detailed description of the moon, this site includes statistics and information about the surface, eclipses, and phases of the moon, along with details of the U.S. *Apollo* landing missions—all supported by a good selection of images.

Mount Stromlo and Siding Spring Observatories
http://msowww.anu.edu.au/

This is a searchable site of Australia's leading observatories. In addition to scientific data of interest only to astronomers, there is general information about the observatories and the work being conducted there, along with downloadable images of space.

Mullard Radio Astronomy Observatory (MRAO)
http://www.mrao.cam.ac.uk/

Site of Cambridge University's radio observatory. Contents include a description of current research interests, technical details of the four telescopes at the English site, details of MRAO publications, links to individual scientists, and details of their research projects.

NASA Home Page
http://www.nasa.gov/

This site offers the latest news from NASA, plus the most recent images from the Hubble Space Telescope, answers to questions about NASA resources and the space program, and a gallery of video, audio clips, and still images.

NASA Shuttle Web
http://spaceflight.nasa.gov/index.html

This is the official NASA site for all U.S. shuttle missions. Here you will find comprehensive technical and nontechnical information, both textual and graphic, and an extensive list (with answers) of frequently asked questions. You can also send questions to shuttle crew members during missions. There are plain-English explanations of NASA's bewildering jargon and acronyms and helpful links to related sites.

National Astronomy and Ionosphere Center (NAIC)
http://www.naic.edu/

This site offers reports from the world's largest radio telescope, at Arecibo, Puerto Rico. It includes readily understandable information about the work of the NAIC and facilities for visitors and a wealth of data for professional radio astronomers, a newsletter reporting the latest research, a gallery of photos, and links to a number of related sites.

Neptune
http://www.hawastsoc.org/solar/eng/neptune.htm

The detailed description of the planet Neptune at this site includes a chronology of the exploration of the planet, along with statistics and information on its rings, moons, and satellites—all supported by a good selection of images.

Nine Planets
http://seds.lpl.arizona.edu/nineplanets/nineplanets/nineplanets.html

This site presents a multimedia tour of the solar system, with descriptions of each of the planets and major moons and appendices on such topics as astronomical names and how they are assigned, the origin of the solar system, and hypothetical planets.

Nuffield Radio Astronomy Laboratories
http://www.jb.man.ac.uk/

This site provides information about the work of Britain's premier radio telescope facility, at Jodrell Bank, Cheshire, England. There is a full description of current research, a listing of astronomers, and details for visitors. An intimate feel of the world of radio astronomy is provided by the link to the JB Alternative Page, on which students report on their work and post their favorite images of space.

Parkes Radio Telescope Home Page
http://www.pks.atnf.csiro.au/

A guide to the work of Australia's main radio telescope, this site includes scientific details for radio astronomers, information of general interest, and education projects for school students and visitors. You can access astronomical images, including an interactive map of the radio sky, and get daily updated information on what the radio astronomers are doing.

Pluto and Charon
http://www.hawastsoc.org/solar/eng/pluto.htm

This site is devoted to our most distant planet and its satellite. It contains a table of statistics, photographs, and an animation of their rotation. You can also find out about NASA's planned mission to Pluto and Charon in 2010.

Practical Guide to Astronomy
http://www.aardvark.on.ca/space/

This well-illustrated guide to astronomy containing explanations of many aspects of the subject, including the Big Bang theories of British theoretical physicist Stephen Hawking, a list of early astronomers and their key discoveries, and an in-depth look at all of the main elements of our solar system.

Project Galileo: Bringing Jupiter to Earth
http://www.jpl.nasa.gov/galileo/

Get full details of the groundbreaking U.S. mission to the solar system's largest planet at this site, which contains regularly updated reports on instructions being given to the probe by staff at the Jet Propulsion Laboratory, Pasadena, California, and of data being received from *Galileo*. There are numerous images of Jupiter and its moons and an animation showing fluctuations in the Great Red Spot. If you cannot find answers to your questions in the list of frequently asked questions, you may e-mail your query to the project.

Project Gemini
http://www.ksc.nasa.gov/history/gemini/gemini.html

This is the official NASA archive of the Gemini space program. There are details (technical and of general interest) on all of the missions included in the project, as well as a comprehensive photo library, a video of a launch, and a search engine.

Project Mercury
http://www.ksc.nasa.gov/history/mercury/mercury.html

This official archive of the program that led to the first manned U.S. space flight—by Alan Shepard in 1961—provides comprehensive details (technical and of general interest) on all of the manned and unmanned flights included in the project, as well as an extensive photo library and a search engine.

Project Skylab
http://www.ksc.nasa.gov/history/skylab/skylab.html

This if the official archive of the project that launched the first U.S. experimental space station, in 1973. It provides comprehensive details (technical and of general interest) on all of the experiments included in the project, a selection of photos and videos, and a search engine.

Rivers of Fire on the Sun
http://www.sciam.com/explorations/091597sun/powell.html

Part of a larger site maintained by the publication *Scientific American*, this page reports on the findings of a team of scientists from Stanford University, California, who identified "rivers" of white-hot plasma flowing on the sun. Color diagrams depict solar flows and labeled cutaway views of the sun's interior. Click on the images to increase their size. The text includes hypertext links to further information, and there is also a list of related links.

Royal Greenwich Observatory (RGO)
http://www.ast.cam.ac.uk/RGO/

This is the official and searchable site of the world's most famous observatory, in Greenwich, England. In addition to a history and a guide for visitors to the observatory's museum, there are comprehensive details of current RGO research (no longer carried out in Greenwich). This is an important site both for students of astronomy and for those seeking information on the latest research.

Saturn
http://www.hawastsoc.org/solar/eng/saturn.htm

How many rings does Saturn have? How many satellites? Find out this and more at this site, which also features a video of a storm in the planet's atmosphere and information on the international *Cassini* mission to Saturn and Titan.

Solar Data Analysis Center Home Page
http://umbra.nascom.nasa.gov/sdac.html

This site includes information about, and images of, the sun taken from space and from ground-based observation posts, as well as numerous links to related pages.

Solar System
http://www.hawastsoc.org/solar/eng/

This educational tour of the solar system contains information and statistics about the sun, Earth, other planets, moons, asteroids, comets, and meteorites found within the solar system, supported by images.

Solar System Live
http://www.fourmilab.ch/solar/solar.html

Take a look at the entire solar system as it might be seen at different times and dates or from different viewpoints.

Stratospheric Observatory for Infrared Astronomy
http://sofia.arc.nasa.gov/

Here you will find general information on the California-based Stratospheric Observatory for Infrared Astronomy (SOFIA) project, run by NASA and the German space agency DARA.

Sun
http://www.hawastsoc.org/solar/eng/sun.htm

Here you will find all you ever wanted to know about our closest star, including cross sections, photographs, a history of exploration, animations of eclipses, and much more. You can also take a multimedia tour of the sun and find out what the current day's weather is like there.

Sun and Moon Data for One Day
http://aa.usno.navy.mil/AA/data/docs/RS_OneDay.html

Part of a larger site on astronomical data maintained by the U.S. Naval Observatory, Washington, D.C., this site includes the times of sunrise, sunset, moonrise, and moonset, transits of the sun and moon, and the beginning and end of twilight (that period when natural light is provided by the upper atmosphere, which receives direct sunlight and reflects part of it toward Earth's surface), along with information on the moon's phase. You can access information by filling out one of two forms, depending on whether you live in or outside the United States. For U.S. cities and towns, there are links to U.S. Census Bureau maps of the area for which astronomical data has been given. There are also sections on frequently asked questions and research information.

Sunspots
http://athena.wednet.edu/curric/space/sun/sunspot.html

This site offers a well-written and easily understandable explanation of sunspots. Features include a high-resolution image of a group of sunspots and a link for further sunspot pictures and information on how to observe sunspots and the sun's cycle of magnetic activity. For more detailed information, there are links to solar observatories.

Uranus
http://www.hawastsoc.org/solar/eng/uranus.htm

Did you know that Uranus is tipped on its side? Find out more about Uranus, its rings, and its moons at this site. Also included are a table of statistics about the planet, photographs, and animations of it rotating.

Virtual Mars
http://members.aol.com/edhobbs/applets/vmars/

View a 3-D image of Mars. Clicking on a land feature centers your view at that spot and allows you to rotate the planet. Click on the button below the image to toggle the names of surface features and landing sites, including the U.S. *Pathfinder* and *Viking 2* sites. Night and day areas are recalculated every five minutes to adjust to rotation.

Voyager Project Home Page
http://vraptor.jpl.nasa.gov/voyager/voyager.html

This site provides comprehensive information on the U.S. *Voyager* probes and what they have told us about the farther reaches of the solar system. There are details of the organization of the *Voyager* mission, the systems that are still functioning, and the data they are sending back—and a host of amazing facts in the Gee-Whiz section.

Welcome to the Mars Missions, Year 2000 and Beyond!
http://marsweb.jpl.nasa.gov/

> This well-presented site offers comprehensive information on current and future U.S. missions to Mars. There are fascinating and well-written accounts of the *Pathfinder* and *Global Surveyor* probes, and large numbers of images of the Red Planet.

WITS Simulation of the Mars Pathfinder Sojourner Rover
http://mars.graham.com/wits/

> Developed by NASA's Jet Propulsion Laboratory in Pasadena, California, this site allows you to interact with a simulated rover, telling it where to go and what to do. The images you see are the actual images *Pathfinder* beamed back to Earth from Mars during its 1998 mission. Other features include an online tutorial and documentation.

Innovations in Astronomy

Part II

7

Dictionary of Terms and Concepts

aberration of starlight

Apparent displacement of a star from its true position, due to the combined effects of the speed of light and the speed of Earth in orbit around the sun (about 30 kps/18.5 mps). Aberration, discovered in 1728 by English astronomer James Bradley, was the first observational proof that Earth orbits the sun.

During a year, the apparent position of a star describes a curve around its true position. This curve is an ellipse except when the star lies on the ecliptic, when it is a line along which the star appears to move backward and forward, and when it is in the pole of the ecliptic, the curve being then a circle. As aberration depends upon the ratio between the velocity of light and the velocity of Earth, the determination of this constant of aberration provides a method of calculating an approximate value of one of these velocities if the other is known.

absolute magnitude

A measure of the intrinsic brightness of a celestial body in contrast to its apparent brightness or magnitude as seen from Earth. For a non-self-luminous body like an asteroid, the absolute magnitude is the magnitude it would appear to have if it were one astronomical unit (AU, or 149.6 milion km/92.96 million mi) from both the sun and Earth with the phase angle zero. For a self-luminous body like a star or a galaxy, the absolute magnitude is the magnitude it would appear to have if it were at a distance of 10 parsecs (or 32.616 light-years).

absorption lines

Dark line in the spectrum of a hot object due to the presence of absorbing material along the line of sight. Absorption lines are caused by atoms absorbing light from the source at sharply defined wavelengths. Numerous absorption lines in the spectrum of the sun (Fraunhofer lines) allow astronomers to study the composition of the sun's outer layers. Absorption lines in the spectra of stars give clues to the composition of interstellar gas.

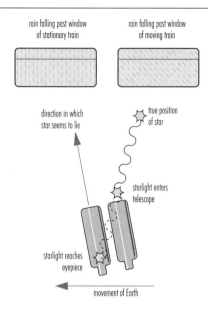

rain falling past window
of stationary train

rain falling past window
of moving train

direction in which
star seems to lie

true position
of star

starlight enters
telescope

starlight reaches
eyepiece

movement of Earth

aberration of starlight

acceleration, secular

The continuous and nonperiodic change in orbital velocity of one body around another, or the axial-rotation period of a body. An example is the axial rotation of Earth. This is gradually slowing down owing to the gravitational effects of the moon and the resulting production of tides, which have a frictional effect on Earth. However, the angular momentum of the Earth-moon system is maintained, because the momentum lost by Earth is passed to the moon. This results in an increase in the moon's orbital period and a consequential moving away from Earth. The overall effect is that Earth's axial-rotation period is increasing by about fifteen-millionths of a second per year, and the moon is receding from Earth at about 4 cm/1.5 in a year.

accretion

In astrophysics, a process by which an object gathers up surrounding material by gravitational attraction, thus simultaneously increasing in mass and releasing gravitational energy. Accretion onto compact objects, such as white dwarfs, neutron stars, and black holes, can release large amounts of gravitational energy and is believed to be the power source for active galaxies. Accreted material falling toward a star may form a swirling disk of material, known as an accretion disk, that can be a source of X-rays.

accretion disk

A flattened ring of gas and dust orbiting an object in space, such as a star or a black hole. The orbiting material is accreted (gathered in) from a neighboring object, such as another star. Giant accretion disks are thought to exist at the centers of some galaxies and quasars. If the central object of the accretion disk has a strong gravitational field, as does a neutron star or a black hole, gas falling onto the accretion disk releases energy, which heats the gas to extreme temperatures and emits short-wavelength radiation, such as X-rays.

Achernar, or Alpha Eridani

Brightest star in the constellation Eridanus and the ninth-brightest star in the sky. It is a hot, luminous, blue star with a true luminosity 250 times that of the sun. It is 144 light-years away from Earth.

achondrite

Type of meteorite. They account for about 15 percent of all meteorites and lack the chondrules (silicate spheres) found in chondrites.

Acrux, or Alpha Crucis

Brightest star in the constellation group Crux and the thirteenth brightest star in the sky. It is a double star 360 light-years from Earth.

active galaxy

A type of galaxy that emits vast quantities of energy from a small region at its center called the active galactic nucleus (AGN). Active galaxies are subdivided into radio galaxies, Seyfert galaxies, BL Lacertae objects, and quasars.

Active galaxies are thought to contain black holes with a mass 10^8 times that of the sun, drawing stars and interstellar gas toward them in a process of accretion. The gravitational energy released by the in-falling material is the power source for the AGN. Some of the energy may appear as a pair of opposed jets emerging from the nucleus. The orientation of the jets to the line of sight and their interaction with surrounding material determine the type of active galaxy that is seen by observers.

ADEOS

See Advanced Earth Observing Satellite (ADEOS).

Adhara

Star 600 light-years from Earth. It plays a greater part in ionizing hydrogen (i.e., producing electrically charged hydrogen atoms) in our region of the galaxy than all of the other three million stars lying closer to our local cloud, according to U.S. astronomers. Most of space contains unionized hydrogen in tiny amounts, but Adhara is connected to our local cloud by a tunnel of almost hydrogen-free space. This means its ionizing radiation reaches our local cloud without obstacle. The surface temperature of Adhara is 21,000 K (20,100°C/37,292°F), almost four times that of the sun; apart from the sun, it is the brightest source of extreme ultraviolet (UV) radiation reaching Earth.

Advanced Earth Observing Satellite (ADEOS)

Japanese remote-sensing satellite launched in August 1996. It gathers data on climate change, the environment, and Earth and ocean processes. ADEOS is carrying eight instruments: Ocean Color and Temperature Scanner (OCTS); Advanced Visible and Near-Infrared Radiometer (AVNIR), Interferometric Monitor for Greenhouse Gases, Improved Limb Atmospheric Spectrometer (ILAS), Retroreflector in Space (RIS), Scatterometer (NSCAT), Total Ozone Mapping Spectrometer (TOMS), and Polarization and Directionality of the Earth's Reflectances (POLDER). NSCAT and TOMS are instruments of the U.S. National Aeronautics and Space Administration (NASA); POLDER is French; and the rest are Japanese.

albedo

The fraction of the incoming light reflected by a body, such as a planet. A body with a high albedo, near 1, is very bright, while a body with a low albedo, near 0, is dark. The moon has an average albedo of 0.12, Venus 0.76, and Earth 0.37.

Alcyone, or Eta Tauri

Brightest member of the Pleiades cluster, a third-magnitude blue star. It is one of the cluster of stars in the zodiacal constellation of Taurus.

Aldebaran, or Alpha Tauri

Brightest star in the constellation Taurus and the fourteenth-brightest star in the sky; it marks the eye of the "bull." Aldebaran is a red giant 65 light-years away from Earth, shining with a true luminosity of about one hundred times that of the sun.

Alfonsine tables

Medieval astronomical work, giving the positions of the planets and other data, prepared during the thirteenth-century reign of the Spanish king Alfonso X ("el Sabio") of Castile and León. The Alfonsine tables were the most important astronomical tables until the sixteenth century.

Algol, or Beta Persei

Eclipsing binary, a pair of orbiting stars in the constellation Perseus, one of which eclipses the other every sixty-nine hours, causing its brightness to drop by two-thirds. This was the first such eclipsing binary to be recognized. Later, more refined observations showed that there was a third, fainter star that revolved round the brighter pair every 1.87 years.

Algonquin Radio Observatory

Site in Ontario, Canada, of the radio telescope, 46 m/150 ft in diameter, of the National Research Council of Canada, which opened in 1966.

almucantar

Small circle on the celestial sphere parallel to the horizon; a circle of altitude.

Alpha Aquilae
See Altair, or Alpha Aquilae.

Alpha Aurigae
See Capella, or Alpha Aurigae.

Alpha Boötis
See Arcturus, or Alpha Boötis.

Alpha Canis Majoris
See Sirius, or Dog Star, or Alpha Canis Majoris.

Alpha Canis Minoris
See Procyon, or Alpha Canis Minoris.

Alpha Canum Venaticorum
See Cor Caroli, or Alpha Canum Venaticorum.

Alpha Carinae
See Canopus, or Alpha Carinae.

Alpha Centauri, or Rigil Kent
Brightest star in the constellation Centaurus, the third-brightest star in the sky, and 4.3 light-years away from Earth. It is actually a triple star; the two brighter stars orbit each other every eighty years, and the third, Proxima Centauri, is the closest star to the sun, 4.2 light-years away from Earth, 0.1 light-years closer than the other two.

Alpha Crucis
See Acrux, or Alpha Crucis.

Alpha Cygni
See Deneb, or Alpha Cygni.

Alpha Eridani
See Achernar, or Alpha Eridani.

Alpha Geminorum
See Achernar, or Alpha Eridani.

Alpha Leonis
See Regulus, or Alpha Leonis.

Alpha Lyrae
See Vega, or Alpha Lyrae.

Alpha Orionis
See Betelgeuse, or Alpha Orionis.

Alpha Piscis Austrini
See Fomalhaut, or Alpha Piscis Austrini.

Alpha Scorpii
See Antares, or Alpha Scorpii.

Alpha Tauri
See Aldebaran, or Alpha Tauri.

Alpha Virginis
See Spica, or Alpha Virginis.

Alphonsus
Name given to a walled plain, 110 km/70 mi in diameter, a little south of the center of the moon's disk. The Soviet astronomer N. A. Kozyrev speculated in 1958 that the central peak of Alphonsus was the site of a volcanic eruption, which he observed spectroscopically, but his interpretation has been disputed.

Alps, Lunar
Conspicuous mountain range on the moon, northeast of the Sea of Showers (Mare Imbrium), cut by a valley 150 km/93 mi long. The highest peak is Mont Blanc, about 3,660 m/12,000 ft.

Altair, or Alpha Aquilae
Brightest star in the constellation Aquila and the twelfth-brightest star in the sky. It is a white star about 16 light-years away from Earth and forms the so-called Summer Triangle with the stars Deneb (in the constellation Cygnus) and Vega (in Lyra).

altazimuth
Astronomical instrument designed for observing the altitude and azimuth of a celestial object. It is essentially a large precision theodolite, i.e., an instrument for the measurement of horizontal and vertical angles.

altitude, or elevation
The angular distance of an object above the horizon, ranging from 0° on the horizon to 90° at the zenith. Together with azimuth, it forms the system of horizontal coordinates for specifying the positions of celestial bodies.

Ames Research Center
Space-research installation of the U.S. National Aeronautics and Space Administration (NASA) at Mountain View, California, for the study of aeronautics and

life sciences. It has managed the Pioneer series of planetary probes and is involved in the search for extraterrestrial life.

Andromeda

Major constellation of the Northern Hemisphere, visible in the fall. Its main feature is the Andromeda galaxy. The star Alpha Andromedae forms one corner of the Square of Pegasus. It is named for the princess of Greek mythology.

Andromeda galaxy

Galaxy 2.2 million light-years from Earth in the constellation Andromeda and the most distant object visible to the naked eye. It is the largest member of the Local Group of galaxies. Like our Milky Way, it is a spiral orbited by several companion galaxies, but it contains about twice as many stars as the Milky Way. It is about 200,000 light-years across. In 1993, U.S. astronomers detected two components at the center of Andromeda, indicating that it may have a double nucleus. The fainter of the two components lies at the exact center of Andromeda, while the rest of the galaxy orbits it. A black hole with a mass tens of millions times that of the sun has been detected here. The other, brighter component of the "double nucleus" may be a remnant of a galaxy that collided with Andromeda, although it was not fully understood by the end of the twentieth century.

Anglo-Australian Telescope

Large telescope on Siding Spring Mountain, New South Wales, Australia, part of the Australia Telescope National Facility run by the Commonwealth Scientific and Industrial Research Organization (CSIRO).

annular eclipse

Solar eclipse in which the moon does not completely obscure the sun, and a thin ring of sunlight remains visible. Annular eclipses occur when the moon is at its farthest point from Earth.

Antares, or Alpha Scorpii

Brightest star in the constellation Scorpius and the fifteenth-brightest star in the sky. It is a red supergiant several hundred times larger than the sun and perhaps ten thousand times as luminous, lies about 420 light-years away from Earth, and fluctuates slightly in brightness.

anthropic principle

In science, the idea that the universe is the way it is because, if it were different, we would not be here to observe it. The principle arises from the observation that, if the laws of science were even slightly different, it would have been impossible for intelligent life to evolve. For example, if the electric charge on the electron were only slightly different, stars would have been unable to burn hydrogen and produce the chemical elements that make up our bodies. Scientists are undecided whether the principle is an insight into the nature of the universe or a piece of circular reasoning.

Antlia
A small, faint constellation in the Southern Hemisphere identified with an air pump.

Apache Point Observatory
U.S. observatory in the Sacramento Mountains of New Mexico containing a 3.5-m/138-in reflector; opened in 1990, it is operated by the Astrophysical Research Consortium (the Universities of Washington, Chicago, Princeton, New Mexico, and Washington State).

apastron
The point at which an object traveling in an elliptical orbit around a star is at its farthest from the star. The term is usually applied to the position of the minor component of a binary star in relation to the primary component. Its opposite is periastron.

Apennines, Lunar
Mountain range on the moon, southeast of the Sea of Showers (Mare Imbrium). It is 740 km/460 mi long, and the highest summit, Huyghens, is 5,500 m/18,045 ft.

aperture synthesis
A technique used in radio astronomy in which several small radio dishes are linked together to simulate the performance of one very large radio telescope, which can be many kilometers in diameter.

aphelion
The point at which an object, traveling in an elliptical orbit around the sun, is at its farthest from the sun. The Earth is at its aphelion on July 5.

apogee
The point at which an object, traveling in an elliptical orbit around Earth, is at its farthest from Earth.

Apollo asteroid
Member of a group of asteroids whose orbits cross that of Earth. They are named for the first of their kind, Apollo, discovered in 1932 by German astronomer Karl Reinmuth, and then lost until 1973. Apollo asteroids are so small and faint that they are difficult to see except when close to Earth (Apollo is about 2 km/1.2 mi across).

Apollo asteroids can collide with Earth from time to time. In December 1994, the Apollo asteroid 1994 XM1 passed 100,000 km/60,000 mi from Earth, the closest observed approach of any asteroid. A collision with an Apollo asteroid sixty-five million years ago has been postulated as one of the causes of the extinction of the dinosaurs. A closely related group, the Amor asteroids, come close to Earth but do not cross its orbit.

Apollo Project

U.S. space project to land a person on the moon, achieved July 20, 1969, when Neil Armstrong was the first to set foot there. He was accompanied on the moon's surface by "Buzz" Aldrin; Michael Collins remained in the orbiting command module.

The program was announced in 1961 by President John F. Kennedy. The world's most powerful rocket, *Saturn V,* was built to launch the *Apollo* spacecraft, which carried three astronauts. When the spacecraft was in orbit around the moon, two astronauts would descend to the surface in a lunar module to take samples of rock and set up experiments that would send data back to Earth. After three other preparatory flights, *Apollo 11* made the first lunar landing. Five more crewed landings followed, the last in 1972. The total cost of the program was more than $24 billion.

Apollo 1
During a preliminary check on the ground, the three crew members were killed by a fire January 27, 1967.

Apollo 4
Launched November 9, 1967, into an orbit around Earth; the first time the *Saturn V* rocket was used.

Apollo 7
The first Apollo mission carrying a crew; *Apollo 7* was a test flight sent into orbit around Earth October 11, 1968.

Apollo 8
Launched December 21, 1968; it was the first rocket to take a crew around the moon.

Apollo 9
Launched March 3, 1969; the lunar module was tested in orbit around Earth.

Apollo 10
Launched May 18, 1969; the lunar module was successfully tested 14.5 km/9 mi above the surface of the moon.

Apollo 11
Launched July 16, 1969; Armstrong and Aldrin landed the lunar module (named *Eagle*) in an area called the Sea of Tranquility on the moon's surface July 20, 1969. Armstrong had to land manually because the automatic navigation system was heading for a field of boulders. On landing, Armstrong announced: "Tranquility base here. The Eagle has landed." The module remained on the moon for 21.6 hours, during which time the astronauts collected rocks,

set up experiments, and mounted a U.S. flag. Apart from a slight wobble when rejoining the command module, the return flight went without a hitch. After splashdown, the astronauts were quarantined as a precaution against unknown illnesses from the moon.

Apollo 12
Launched November 14, 1969; in spite of twice being struck by lightning, it achieved another successful moon landing.

Apollo 13
Intended to be the third moon landing, *Apollo 13* was launched April 11, 1970, with the crew of John Swigert, Fred Haise, and James Lovell. On the third day of the mission, Swigert reported to mission control in Houston, Texas: "We seem to have a problem." An electrical fault had caused an explosion in one of the oxygen tanks, cutting off supplies of power and oxygen to the command module. The planned landing was abandoned, and the rocket was sent round the moon before heading back to Earth. The crew used the lunar module *Aquarius* as a "lifeboat," though they had to endure near-freezing temperatures to save power, making sleep almost impossible. Attempting reentry in the crippled ship almost led to disaster, but the crew splashed down safely April 17.

Apollo 14
Launched January 31, 1971; under the command of U.S. astronaut Alan Shepard, it reaches the moon on February 5 and returns to Earth on February 8 with samples of lunar rock.

Apollo 15
Launched July 26, 1971; on its mission, the first surface vehicle was used on the moon, the Lunar Rover.

Apollo 16
Launched April 16, 1972; the mission gathers lunar soil and rock during a record seventy-one hours and two minutes on the moon.

Apollo 17
Launched December 7, 1972; the last of the Apollo moon landings. Detailed geological studies were carried out, and large amounts of rock and soil were brought back.

Apollo-Soyuz test project
Joint U.S.-Soviet space mission in which an *Apollo* and a *Soyuz* craft docked while in orbit around Earth on July 17, 1975. The craft remained attached for two days, and crew members were able to move from one craft to the other through an air lock attached to the nose of the *Apollo*. The mission was designed to test rescue procedures and also had political significance.

apparent

In astronomy, term synonymous with "observed." It is used, for example, in "apparent magnitude "and "apparent star place"—the observed position, which will yield the mean position given in star catalogs only when it has been corrected for the effects of aberration, precession, and nutation.

apsis (plural apsides)

Either of two points in an orbit, one being at maximum and the other at minimum distance from the controlling central body. The apogee and the perigee of the moon are apsides, as are the aphelion and the perihelion of a planet.

Apus

Faint constellation in the souuthern sky near the south pole. Its name derives from the Greek for Bird of Paradise.

Aquarius

Zodiacal constellation a little south of the celestial equator near Pegasus. Aquarius is represented as a man pouring water from a jar; the constellation is popularly called the "water bearer." The sun passes through Aquarius from late February to early March. In astrology, the dates for Aquarius, the eleventh sign of the zodiac, are about January 20–February 18.

Aquila

Constellation on the celestial equator. Its brightest star is first-magnitude Altair, flanked by the stars Beta and Gamma Aquilae. It is represented by an eagle. Nova Aquilae, which appeared in June 1918, shone for a few days nearly as brightly as Sirius.

Ara

Small, faint constellation in the southern sky named for the Latin word for altar. Among those listed by Greek astronomer Ptolemy (second century A.D.).

arc minute, arc second

Units for measuring small angles, used in geometry, surveying, mapmaking, and astronomy. An arc minute (symbol ') is one-sixtieth of a degree, and an arc second (symbol ") is one-sixtieth of an arc minute. Small distances in the sky, as between two close stars or the apparent width of a planet's disk, are expressed in minutes and seconds of arc.

Arcturus, or Alpha Boötis

Brightest star in the constellation Boötes and the fourth-brightest star in the sky. Arcturus is a red giant twenty-eight times larger than the sun and seventy times more luminous; it is 36 light-years from Earth.

Its name is derived from the Greek and signifies "the guardian of the bear." As the brightest star in the northern celestial hemisphere, it must always have been conspicuous, but some of the ancient literary references to it, such as those in the biblical Book of Job, are probably intended to include the Great Bear as well.

Arcturus was the first star for which a proper motion was detected. Edmond Halley, in 1718, noticed that, relative to the surrounding stars, it had moved by about 1° from the position recorded by Greek astronomer Ptolemy in the second century A.D. in the *Almagest*.

Arecibo Observatory

Site in Puerto Rico of the world's largest single-dish radio telescope, 305 m/1,000 ft in diameter. It is built in a natural hollow and uses the rotation of Earth to scan the sky. It has been used for radar work on the planets and for conventional radio astronomy and is operated by the National Astronomy and Ionosphere Center of Cornell University, Ithaca, New York, and the National Science Foundation.

In 1996, it received a $25 million upgrade, increasing the sensitivity of the disk tenfold. Two new mirrors were also added, and the observation frequency was increased from 3,000 megahertz to up to 10,000 megahertz. Another upgrade took place 1997, when a new receiver capable of monitoring 168 million radio channels, SERENDIP IV, was added to the facility.

Argo

Very large bright constellation of the Southern Hemisphere, represented as a ship. It was described by Greek astronomer Ptolemy in the second century A.D. but now, to avoid confusion, is split into four: Carina, the keel; Puppis, the stern; Vela, the sail; and Pyxis, the compass. The last was formed from stars in Malus, the mast, another former subdivision of Argo that has not survived as a separate constellation.

argument

In astronomy, a term used in a particular sense in orbit theory. For example, in "the argument of perihelion," it is used for the angle between the ascending node (the point at which the orbit intersects the plane of the ecliptic, moving from south to north of the ecliptic) and the perihelion.

Ariane

Launch vehicle built in a series by the European Space Agency; its first flight was in 1979. The launch site is at Kourou, French Guiana. *Ariane* is a three-stage rocket using liquid fuels. Small solid-fuel and liquid-fuel boosters can be attached to its first stage to increase carrying power.

Since 1984, it has been operated commercially by Arianespace, a private company financed by European banks and aerospace industries. A more powerful version, *Ariane 5*, was launched June 4, 1996, and was intended to carry astronauts aboard the *Hermes* space plane. However, it went off course immediately after takeoff, turned on its side, broke in two, and disintegrated. A fault in the software controlling the takeoff trajectory was to blame. A mostly successful test flight for *Ariane 5* was completed in November 1997.

Ariel

Series of six British satellites launched by the United States (1962–1979), the most significant of which was *Ariel 5* in 1974, which made a pioneering survey of the sky at X-ray wavelengths.

Ariel

The inmost of the five major moons (or satellites) of the planet Uranus. Ariel is thought to consist primarily of water ice; it has an estimated diameter of 1,330 km/825 mi and orbits at a distance of 191,020 km/118,432 mi from the center of Uranus.

Aries

Zodiacal constellation in the Northern Hemisphere between Pisces and Taurus, near Auriga, represented as the legendary ram whose golden fleece was sought by Jason and the Argonauts. Its most distinctive feature is a curve of three stars of decreasing brightness. The brightest of these is Hamal, or Alpha Arietis, 65 light-years from Earth.

The sun passes through Aries from late April to mid-May. In astrology, the dates for Aries, the first sign of the zodiac, are about March 21–April 19. The spring equinox once lay in Aries but has now moved into Pisces through the effect of Earth's precession (wobble).

Aristarchus

Crater, 80 km/50 mi in diameter, in the northeast quadrant of the moon. It is very bright and remains visible as a luminous spot after all of the surrounding region is in shadow.

armillary sphere

Earliest-known astronomical device, in use from the third century B.C. It showed Earth at the center of the universe, surrounded by a number of movable metal rings representing the sun, the moon, and the planets. The armillary sphere was originally used to observe the heavens and later for teaching navigators about the arrangements and movements of the heavenly bodies.

ashen light

A faint glow occasionally reported in the dark hemisphere of Venus when the planet is in a crescent phase. Its origin is unknown, but it may be related to the terrestrial airglow, i.e., the faint and variable light caused by interaction of high-energy solar radiation with the upper atmosphere of Earth.

asterism

A group of stars, not necessarily a complete constellation, forming an easily recognizable figure, such as the Big Dipper (part of Ursa Major), the Pleiades, or Orion's belt.

asteroid, or minor planet

Any of many thousands of small bodies, composed of rock and iron, that orbit the sun. Most lie in a belt between the orbits of Mars and Jupiter and are thought to be fragments left over from the formation of the solar system. About 100,000 may exist, but their total mass is only a few hundredths the mass of the moon.

They include Ceres (the largest asteroid, 940 km/584 mi in diameter), Vesta (which has a light-colored surface and is the brightest as seen from Earth), Eros, and Icarus. Some asteroids are in orbits that bring them close to Earth, and some, such as the Apollo asteroids, even cross Earth's orbit; at least some of these may be remnants of former comets. One group, the Trojans, moves along the same orbit as Jupiter, 60° ahead and behind the planet. One unusual asteroid, Chiron, orbits beyond Saturn.

In February 1996, the U.S. National Aeronautics and Space Administration (NASA) launched the *Near Earth Asteroid Rendezvous* (NEAR) spacecraft to study Eros in order to ascertain what asteroids are made of and whether they are similar in structure to meteorites. In 1997, it flew past the asteroid Mathilde, revealing a 25-km/15.6-mi crater covering the 53-km/33-mi asteroid. The Near Earth Asteroid Tracking (NEAT) system had detected more than ten thousand asteroids by August 1997.

The first asteroid was discovered January 1, 1801, by the Italian astronomer Giuseppe Piazzi at the Palermo Observatory, Sicily. The first asteroid moon was observed in 1993 by the space probe *Galileo* orbiting the asteroid Ida.

Bifurcated asteroids, first discovered in 1990, are, in fact, two chunks of rock that touch each other. It may be that at least 10 percent of asteroids approaching Earth are bifurcated.

astrometry

Measurement of the precise positions of stars, planets, and other bodies in space. Such information is needed for practical purposes, including accurate timekeeping, surveying and navigation, and calculating orbits and measuring distances in space. Astrometry is not concerned with the surface features or the physical nature of the body under study.

Before telescopes, astronomical observations were simple astrometry. Precise astrometry has shown that stars are not fixed in position but have a proper motion caused as they and the sun orbit the Milky Way galaxy. The nearest stars also show parallax (apparent change in position), from which their distances can be calculated. Above the distorting effects of the atmosphere, satellites such as *Hipparcos*, launched by the European Space Agency in 1989, can make even more precise measurements than ground telescopes, and so refine the distance scale of space.

astronaut

Person making flights into space; the term "cosmonaut" is used in the West for any astronaut from the former Soviet Union.

astronautics
Science of space travel.

astronomical unit
Unit (symbol AU) equal to the mean distance of Earth from the sun: 149.6 million km/92.96 million mi. It is used to describe planetary distances. Light travels this distance in approximately 8.3 minutes.

astrophotography
Use of photography in astronomical research. The first successful photograph of a celestial object was the daguerreotype plate of the moon taken in March 1840 by English scientist John W. Draper of the United States. The first photograph of a star, Vega, was taken in 1850 by U.S. astronomer William Bond. Modern-day astrophotography uses techniques such as charge-coupled devices (CCDs).

Before the development of photography, observations were gathered in the form of sketches made at the telescope. Several successful daguerreotypes were obtained prior to the introduction of wet-plate collodion about 1850. The availability of this more convenient method allowed photography to be used on a more systematic basis, including the monitoring of sunspot activity. Dry plates were introduced in the 1870s, and, in 1880, Henry Draper obtained a photograph of the Orion Nebula. The first successful image of a comet was obtained in 1882 by the Scottish astronomer David Gill, his plate displaying excellent star images. Following this, Gill and Dutch astronomer J. C. Kapteyn compiled the first photographic atlas of the southern sky, cataloging almost half a million stars.

Modern-day electronic innovations, notably charge-coupled devices (CCDs), provide a more efficient light-gathering capability than photographic film and enable information to be transferred to a computer for analysis. However, CCD images are expensive and very small in size compared to photographic plates. Photographic plates are better suited to wide-field images, whereas CCDs are used for individual objects, which may be very faint, within a narrow field of sky.

astrophysics
Study of the physical nature of stars, galaxies, and the universe. It began with the development of spectroscopy in the nineteenth century, which allowed astronomers to analyze the composition of stars from their light. Astrophysicists view the universe as a vast natural laboratory in which they can study matter under conditions of temperature, pressure, and density that are unattainable on Earth.

Atlas rocket
U.S. rocket, originally designed and built as an intercontinental ballistic missile (ICBM) but subsequently adapted for space use. *Atlas* rockets launched U.S. astronauts in the Mercury Project into orbit, as well as numerous other satellites and space probes.

Auriga

Constellation of the Northern Hemisphere, represented as a charioteer. Its brightest star is the first-magnitude Capella, about 42 light-years from Earth; Epsilon Aurigae is an eclipsing binary star with a period of twenty-seven years, the longest of its kind (the last eclipse was in 1983).

The charioteer is usually represented as a man holding a bridle in his right hand and supporting a goat and kids on his left arm. The goat is identified with Capella, whose name means "the Little Nanny Goat," and the kids with the three adjacent stars: Epsilon, Eta, and Zeta Aurigae. The charioteer is also identified as Erichthonius, the legendary king of Athens who invented the four-horse chariot.

aurora

Colored light in the night sky near Earth's magnetic poles, called aurora borealis (northern lights) in the Northern Hemisphere and aurora australis (southern lights) in the Southern Hemisphere. Although aurorae are usually restricted to the polar skies, fluctuations in the solar wind occasionally cause them to be visible at lower latitudes. An aurora is usually in the form of a luminous arch, with its apex toward the magnetic pole, followed by arcs, bands, rays, curtains, and coronas, usually green but often showing shades of blue and red and sometimes yellow or white. Aurorae are caused at heights of more than 100 km/60 mi by a fast stream of charged particles from solar flares and low-density "holes" in the sun's corona.

These are guided by Earth's magnetic field toward the north and south magnetic poles, where they enter the upper atmosphere and bombard the gases in the atmosphere, causing them to emit visible light.

AUSSAT

Organization formed in 1981 by the federal government of Australia and Telecom Australia to own and operate Australia's domestic satellite system. The first stage, *Aussat 1,* was launched in 1985 by the U.S. space shuttle *Discovery,* and the third and final stage was launched in 1987 by the European Space Agency's *Ariane* rocket launcher from French Guiana, South America. The AUSSAT satellite system enables people in remote outback areas of Australia to receive television broadcasts.

Australia Telescope National Facility

Giant radio telescope in New South Wales, Australia, operated by the Commonwealth Scientific and Industrial Research Organization (CSIRO). It consists of six 22-m/72-ft antennae at Culgoora, near Narrabri, a similar antenna at Mopra, Siding Spring Mountain, and the 64-m/210-ft Parkes radio telescope—the whole simulating a dish 300 m/186 mi across.

It can be operated in three configurations: (1) the compact array of the six Culgoora antennae, known as the Paul Wild Observatory, which can map the same fine detail as could a telescope 6 km/3.7 mi in diameter; (2) the long baseline array, in which one or more of the antennae in the compact array are linked to the antennae at Siding Spring Mountain and Parkes and which is fifty times better at resolving fine detail than the compact array; and (3) the possibility of linking with other antennae in Australia and overseas, or with the two proposed orbiting radio telescopes scheduled for launch in the 1990s, to form giant arrays that will be up to ten thousand times better at detecting detail than the best existing optical telescopes.

azimuth
The angular distance of an object eastward along the horizon, measured from due north, between the astronomical meridian (the vertical circle passing through the center of the sky and the north and south points on the horizon) and the vertical circle containing the celestial body whose position is to be measured.

Baikonur
Launch site for spacecraft, located at Tyuratam, Kazakhstan, near the Aral Sea; the first satellites and all Soviet space probes and crewed *Soyuz* missions were launched from here. It covers an area of 12,200 km^2/4,675 mi^2, much larger than its U.S. equivalent, the Kennedy Space Center in Florida.

Baily's beads
Bright spots of sunlight seen around the edge of the moon for a few seconds immediately before and after a total eclipse of the sun, caused by sunlight shining between mountains at the moon's edge. Sometimes, one bead is much brighter than the others, producing what is called the diamond-ring effect.

Barnard's Star
Second-closest star to the sun, 6 light-years away from Earth in the constellation Ophiuchus. It is a faint red dwarf of tenth magnitude, visible only through a telescope. It is named for U.S. astronomer Edward E. Barnard, who discovered in 1916 that it has the fastest proper motion of any star, 10.3 arc seconds per year. Some observations suggest that Barnard's Star may be accompanied by planets.

Barringer Crater, or Arizona Meteor Crater, or Coon Butte
Impact crater near Winslow, Arizona, caused by the impact of a 50-m/165-ft iron meteorite about fifty thousand years ago. It is 1.2 km/0.7 mi in diameter and 200 m/660 ft deep, and the walls are raised 50–60 m/165–198 ft above the surrounding desert. It is named for U.S. mining engineer Daniel Barringer, who proposed in 1902 that it was an impact crater rather than a volcanic feature, an idea confirmed in the 1960s by U.S. geologist Eugene Shoemaker.

Bellatrix
Gamma Orionis, a second-magnitude star that marks the western shoulder of Orion.

Beta Centauri
See Hadar, or Beta Centauri.

Beta Crucis
See Mimosa, or Beta Crucis.

Beta Geminorum
See Pollux, or Beta Geminorum.

Beta Orionis
See Rigel, or Beta Orionis.

Beta Persei
See Algol, or Beta Parsei.

Betelgeuse, or Alpha Orionis
Red supergiant star in the constellation of Orion. It is the tenth-brightest star in the night sky, although its brightness varies. It is 1,100 million km/700 million mi across, about eight hundred times larger than the sun, roughly the same size as the orbit of Mars. It is more than ten thousand times as luminous as the sun and lies 310 light-years from Earth. Light takes one hour to travel across the giant star.

Its apparent magnitude varies irregularly between 0.4 and 1.3 in a period of 5.8 years. It was the first star whose angular diameter was measured with the interferometer at the Mount Wilson Observatory, near Los Angeles, California—in 1920. The name is a corruption of the Arabic, describing its position in the shoulder of Orion.

Big Bang
The hypothetical "explosive" event that marked the origin of the universe as we know it. At the time of the Big Bang, the entire universe was squeezed into a hot, superdense state. The Big Bang explosion threw this compact material outward, producing the expanding universe. The cause of the Big Bang is unknown; observations of the current rate of expansion of the universe suggest that it took place between ten billion and twenty billion years ago. The Big Bang theory began modern cosmology.

According to a modified version of the Big Bang, called the inflationary theory, the universe underwent a rapid period of expansion shortly after the Big Bang, which accounts for its current large size and uniform nature. The inflationary theory is supported by the most recent observations of the cosmic background radiation.

Scientists have calculated that one 10^{-36} second (one million-million-million-million-million-millionth of a second) before the Big Bang, the universe was the size of a pea, and the temperature was 10 billion million million million°C/18 billion million million million°F. One second after the Big Bang, the temperature was about 10 billion°C/18 billion°F.

Big Crunch
In cosmology, a possible fate of the universe in which it ultimately collapses to a point following the halting and reversal of the present expansion.

Big Dipper

Name popular in the United States for the seven brightest and most prominent stars in the constellation Ursa Major, which in outline resemble a large dipper.

binary star

Pair of stars moving in orbit around their common center of mass. Observations show that most stars are binary, or even multiple—for example, the nearest star system to the sun, Alpha Centauri.

One of the stars in the binary system Epsilon Aurigae may be the largest star known. Its diameter is twenty-eight hundred times that of the sun. If it were in the position of the sun, it would engulf Mercury, Venus, Earth, Mars, Jupiter, and Saturn. A spectroscopic binary is a binary in which the two stars are so close together that they cannot be seen separately, but their separate light spectra can be distinguished by a spectroscope.

Another type is the eclipsing binary, a double star in which the two stars periodically pass in front of each other as seen from Earth. When one star crosses in front of the other, the total light received on Earth from the two stars declines. The first eclipsing binary to be noticed was Algol in 1670 by Italian astronomer Germiniano Montana.

Alpha Centauri, for example, consists of a star almost identical to the sun with another star about a third as bright closer to it than Neptune is to the sun, i.e., closer than 4.4 billion km/2.794 billion mi. Each of these stars appears to describe an ellipse about the other every eighty years. A third, much fainter star in Alpha Centauri, Proxima Centauri, is too far away to disturb their mutual orbit appreciably. The study of such systems has been rewarding and has provided the only reliable information about the masses of stars. For a few stars, it has also yielded direct measures of their dimensions, shapes, and effective temperatures.

The precision observation of binary stars did not start until the time of German-born British astronomer William Herschel in the eighteenth century, who recorded the most thorough catalog of binary stars. He recorded 848 "double stars" or binaries. Binaries can now be detected in a number of ways: by direct telescopic observation (visual binaries); by suitable interferometers (interferometric binaries); by periodic variations in proper motion (astrometric binaries) and in radial velocity (spectroscopic binaries); and as variable stars (eclipsing binaries).

Each method of observing picks out different samples of binaries. Visual observation selects those of long period, as these are the only ones sufficiently well separated; the interferometer can resolve stars with smaller separations but is limited to those pairs in which the components are nearly equal in brightness; close pairs in which changes of velocity are large and take place in a short period are more likely to be noticed spectroscopically; photometric detection is limited to those systems in which the orbital plane happens to pass close to Earth. The more ways in which a particular binary can be observed, the more detailed the information it can yield about the individual components.

BL Lacertae object

Starlike object that forms the center of a distant galaxy, with a prodigious energy output. BL Lac objects, as they are called, seem to be related quasars and are thought to be the brilliant nuclei of elliptical galaxies. They are so named because the first to be discovered lies in the small constellation Lacerta.

black hole

Object in space whose gravity is so great that nothing can escape from it, not even light. Thought to form when massive stars shrink at the end of their lives, a black hole sucks in more matter, including other stars, from the space around it. Matter that falls into a black hole is squeezed to infinite density at the center of the hole. Black holes can be detected because gas falling toward them becomes so hot that it emits X-rays.

Black holes containing the mass of millions of stars are thought to lie at the centers of quasars. Satellites have detected X-rays from a number of objects that may be black holes, but only four likely black holes in our galaxy had been identified by 1994.

Cygnus X-1, first discovered in 1964 by astronomers at the U.S. Naval Research Laboratory (NRL), is an X-ray source in the constellation of Cygnus. A0620–00, in the constellation of Monoceros, is one of the best black-hole candidates in the galaxy; it was discovered in the 1980s by U.S. astronomers Jeffrey McClintock of the Harvard-Smithsonian Center for Astrophysics and Ronald Remillard of the Massachusetts Institute of Technology. V404 Cygni, close to Cygnus X-1, is a possible black hole discovered in 1992 by Jorge Casares, Phil Charles, and Tim Naylor. Nova Muscae, identified as a black hole in 1992 by McClintock, Remillard, and U.S. astronomer Charles Bailyn of Yale University, New Haven, Connecticut, lies approximately 18,000 light-years from Earth.

In 1997, the Hubble Space Telescope discovered evidence of a black hole 300 million times the mass of the sun. It is located in the middle of galaxy M84 about 50 million light-years from Earth.

Microscopic black holes may have been formed in the chaotic conditions of the Big Bang. British theoretical physicist Stephen Hawking has shown that such tiny black holes could "evaporate" and explode in a flash of energy.

blue shift

A manifestation of the Doppler effect in which an object appears bluer when it is moving toward the observer or the observer is moving toward it (blue light is of a higher frequency than other colors in the spectrum). The blue shift is the opposite of the red shift.

bolometric magnitude

A measure of the brightness of a star over all wavelengths. Bolometric magnitude is related the total radiation output of the star.

Boötes

Constellation of the Northern Hemisphere, represented by a herdsman driving a bear (Ursa Major) around the pole. Its brightest star is Arcturus (or Alpha Boötis), which is 36 light-years from Earth. The herdsman is assisted by the neighboring Canes Venatici, the hunting dogs.

Epsilon Boötis, a double star with blue and yellow components, was called Pulcherrima ("most beautiful") by Russian astronomer Friedrich Wilhelm Struve and was used by German-born British astronomer William Herschel in an unsuccessful attempt to determine a stellar distance.

brown dwarf

An object less massive than a star but heavier than a planet. Brown dwarfs do not have enough mass to ignite nuclear reactions at their centers, but they shine by heat released during their contraction from a gas cloud. Some astronomers believe that vast numbers of brown dwarfs exist throughout the galaxy. Because of the difficulty of detection, none were spotted until 1995, when U.S. astronomers discovered a brown dwarf, GI229B, in the constellation Lepus. It is twenty to forty times as massive as Jupiter but emits only 1 percent of the radiation of the smallest-known star. In 1996, British astronomers discovered four possible brown dwarfs within 150 light-years of the sun.

Caelum

Inconspicuous constellation of the Southern Hemisphere, represented as a sculptor's chisel; it was originally named Caela Sculptoris in 1752 by French astronomer Nicolas Lacaille.

Callisto

Second-largest moon of Jupiter, 4,800 km/3,000 mi in diameter, orbiting every 16.7 Earth days at a distance of 1.9 million km/1.2 million mi from the planet. Its surface is covered with large craters. The U.S. space probe *Galileo* detected molecules containing both carbon and nitrogen atoms on the surface of Callisto, U.S. astronomers announced in March 1997. The presence of carbon and nitrogen atoms may indicate that Callisto harbored life at some time.

Camelopardalis

Faint constellation of the north polar region, with a long, straggling shape, represented as a giraffe.

Cancer

Faintest of the zodiacal constellations (its brightest stars are fourth magnitude). It lies in the Northern Hemisphere between Leo and Gemini and is represented as a crab. The sun passes through the constellation during late July and early August. In astrology, the dates for Cancer are about June 22–July 22. Cancer's most distinctive feature is the open star cluster Praesepe, popularly known as the Beehive, visible to the naked eye as a nebulous patch. In Chaldaean and Platonist philosophy, Cancer was "the Gate of Men," through which souls descended into human bodies, eventually returning to heaven through Capricornus, "the Gate of the Gods."

Canes Venatici
Constellation of the Northern Hemisphere near Ursa Major, identified with the hunting dogs of Boötes, the herder. Its stars are faint, and it contains the Whirlpool galaxy (M51), the first spiral galaxy to be recognized. It contains many objects of telescopic interest, including the relatively bright globular cluster M3. The brightest star, a third-magnitude double, is called Cor Caroli, or Alpha Canum Venaticorum.

Canis Major
Brilliant constellation of the Southern Hemisphere, represented (with Canis Minor) as larger of the two dogs following at the heel of Orion (the other dog being Canis Minor). Its main star, Sirius, is the brightest star in the night sky. Epsilon Canis Majoris is also of the first magnitude, and there are three second-magnitude stars.

Canis Minor
Small constellation along the celestial equator, represented as the smaller of the two dogs of Orion (the other dog being Canis Major). Its brightest star is the first-magnitude Procyon. Procyon and Beta Canis Minoris form what the Arabs called "the Short Cubit," in contrast to "the Long Cubit" formed by Castor and Pollux (Alpha and Beta Geminorum, respectively).

Canopus, or Alpha Carinae
Second-brightest star in the night sky (after Sirius), lying in the southern constellation Carina. It is a first-magnitude yellow-white supergiant about 120 light-years from Earth and thousands of times more luminous than the sun.

Cape Canaveral
Promontory on the Atlantic coast of Florida, 367 km/228 mi north of Miami, used as a rocket-launch site by the U.S. National Aeronautics and Space Administration (NASA). It was known as Cape Kennedy (1963–1973), for former president John F. Kennedy. The Kennedy Space Center is nearby.

Capella, or Alpha Aurigae
Brightest star in the constellation Auriga and the sixth-brightest star in the night sky. It is a visual and spectroscopic binary that consists of a pair of yellow-giant stars 42 light-years from Earth, orbiting each other every 104 days. It is a first-magnitude star whose Latin name means "the Little Nanny Goat"; its kids are the three adjacent stars Epsilon, Eta, and Zeta Aurigae.

Capricorn
See Capricornus.

Capricornus
Zodiacal constellation in the Southern Hemisphere next to Sagittarius. It is represented as a sea-goat, and its brightest stars are third magnitude. The sun passes through it late January to mid-February. In astrology, the dates for Capricornus (popularly known as Capricorn) are about December 22–January 19.

In Chaldaean and Platonist philosophy, Capricornus was regarded as the "Gate of the Gods" through which souls ascended to heaven, their descent having been through Cancer, "the Gate of Men."

captured rotation, or synchronous rotation
The circumstance in which one body in orbit around another, such as the moon of a planet, rotates on its axis in the same time as it takes to complete one orbit. As a result, the orbiting body keeps one face permanently turned toward the body about which it is orbiting. An example is the rotation of our own moon, which arises because of the tidal effects of Earth over a long period of time.

carbon cycle
In astrophysics, a sequence of nuclear-fusion reactions in which carbon atoms act as a catalyst to convert four hydrogen atoms into one helium atom with the release of energy. The carbon cycle is the dominant energy source for ordinary stars of mass greater than about 1.5 times the mass of the sun. Nitrogen and oxygen are also involved in the sequence, so it is sometimes known as the carbon-nitrogen-oxygen (CNO) cycle.

Carina
Constellation of the Southern Hemisphere, represented as a ship's keel. Its brightest star is Canopus, the second-brightest in the night sky; it also contains Eta Carinae, a massive and highly luminous star embedded in a gas cloud, perhaps 8,000 light-years away from Earth. Carina was formerly regarded as part of Argo and is situated in one of the brightest parts of the Milky Way.

Cassegrain telescope, or Cassegrain reflector
Type of reflecting telescope in which light collected by a concave primary mirror is reflected onto a convex secondary mirror, which, in turn, directs it back through a hole in the primary mirror to a focus behind it. As a result, the telescope tube can be kept short, allowing equipment for analyzing and recording starlight to be mounted behind the main mirror. All modern large astronomical telescopes are of the Cassegrain type.

It is named for the seventeenth-century French astronomer Cassegrain, who first devised it as an improvement to the simpler Newtonian telescope.

Cassini

Joint space probe of the U.S. National Aeronautics and Space Administration (NASA) and the European Space Agency (ESA) to the planet Saturn. *Cassini* was launched October 15, 1997, to go into orbit around Saturn in 2004, dropping off a subprobe, *Huygens,* to land on Saturn's largest moon, Titan. It was launched on a *Titan 4* rocket, with its electricity supplied by 32 kg/70 lb of plutonium. This is the largest amount of plutonium ever to be sent into space, and it provoked fears of contamination should *Cassini*, or its rocket, malfunction. A *Titan 4* exploded in flight in 1993.

Cassiopeia

Prominent constellation of the Northern Hemisphere, named for the mother of Andromeda. It has a distinctive W-shape and contains one of the most powerful radio sources in the sky, Cassiopeia A. This is the remains of a supernova (star explosion) that occurred ca. A.D. 1702, too far away to be seen from Earth. It was in Cassiopeia that Danish astronomer Tycho Brahe, in 1572, observed a new star, which was probably a supernova since it was visible in daylight and outshone Venus for ten days.

Castor, or Alpha Geminorum

Second-brightest star in the constellation Gemini and the twenty-third-brightest star in the night sky. Along with the brighter Pollux, or Beta Geminorum, it forms a prominent pair at the eastern end of Gemini, representing the head of the twins. Second-magnitude Castor is 45 light-years from Earth and is one of the finest binary stars in the sky for small telescopes. The two main components orbit each other over a period of 467 years. A third, much fainter, star orbits the main pair over a period probably exceeding ten thousand years. Each of the three visible components is a spectroscopic binary, making Castor a sextuple star system.

celestial mechanics

The branch of astronomy that deals with the calculation of the orbits of celestial bodies, their gravitational attractions (such as those that produce Earth's tides), and also the orbits of artificial satellites and space probes. It is based on the laws of motion and gravity laid down by seventeenth-century British physicist Isaac Newton.

celestial sphere

Imaginary sphere surrounding Earth, on which the celestial bodies seem to lie. The positions of bodies such as stars, planets, and galaxies are specified by their coordinates on the celestial sphere. The equivalents of latitude and longitude on the celestial sphere are called declination and right ascension (which is measured in hours from zero to twenty-four). The celestial poles lie directly above Earth's poles, and the celestial equator lies over Earth's equator. The celestial sphere appears to rotate once around Earth each day, actually a result of the rotation of Earth on its axis.

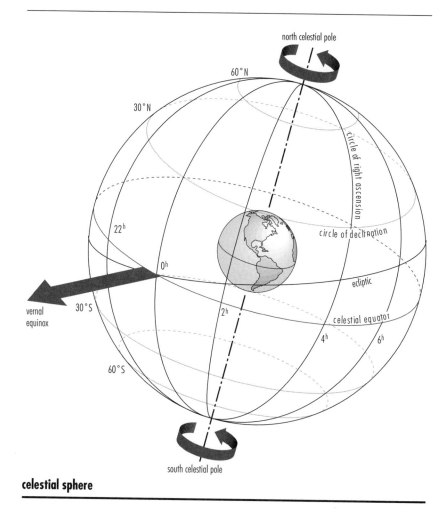

north celestial pole

60°N

30°N

circle of right ascension

22ʰ

circle of declination

0ʰ

ecliptic

vernal
equinox

30°S

2ʰ

celestial equator

4ʰ

6ʰ

60°S

south celestial pole

celestial sphere

centaur

Cometlike object with an unstable orbit of less than two hundred years. Centaurs are 100–400 km/62–250 mi in diameter and are redder than other asteroids. The six known centaurs originated in the Kuiper belt. Chiron and Pholus are centaurs.

Centaurus

Large, bright constellation of the Southern Hemisphere, represented as a centaur. Its brightest star, Alpha Centauri, is a triple star and contains the closest star to the sun, Proxima Centauri, which is only 4.2 light-years away from Earth, 0.1 light-years closer than its companions, Alpha Centauri A and B. Omega Centauri, which is just visible to the naked eye as a hazy patch, is the largest and brightest globular cluster of stars in the sky, 16,000 light-years away from Earth.

Alpha and Beta Centauri are both of the first magnitude and, like Alpha and Beta Ursae Majoris, are known as "the Pointers," as a line joining them leads to Crux. Centaurus A, a galaxy 15 million light-years away from Earth, is a strong source of radio waves and X-rays.

Cepheid variable

Yellow supergiant star that varies regularly in brightness every few days or weeks as a result of pulsations. The time that a Cepheid variable takes to pulsate is directly related to its average brightness; the longer the pulsation period, the brighter the star. This relationship, the Period-Luminosity Law, which was discovered by U.S. astronomer Henrietta Leavitt, allows astronomers to use Cepheid variables as "standard candles" to measure distances in our galaxy and to nearby galaxies.

Cepheus

Constellation of the north polar region, named for King Cepheus of Greek mythology, husband of Cassiopeia and father of Andromeda. It contains the Garnet Star (Mu Cephei), a red supergiant of variable brightness that is one of the reddest-colored stars known, and Delta Cephei, prototype of the Cepheid variables, which are important both as distance indicators and for the information they give about stellar evolution.

Ceres

The largest asteroid, 940 km/584 mi in diameter, and the first to be discovered (by Italian astronomer Giuseppe Piazzi in 1801). Ceres orbits the sun every 4.6 years at an average distance of 414 million km/257 million mi. Its mass is about 0.014 of Earth's moon.

Cerro Tololo Interamerican Observatory

Observatory on Cerro Tololo Mountain in the Chilean Andes operated by AURA (Association of Universities for Research into Astronomy). Its main instrument is a 4-m/158-in reflector, opened in 1974, a twin of that at Kitt Peak, Arizona.

Cetus

Large constellation on the celestial equator, represented as a sea monster or a whale. Cetus contains the long-period variable star Mira, and Tau Ceti, one of the nearest stars, which is visible with the naked eye. It is named for the sea monster sent to devour Andromeda. Mira is sometimes the most conspicuous object in the constellation, but it is more usually invisible to the naked eye.

Challenger

Orbiter used in the U.S. space-shuttle program that, on January 28, 1986, exploded on takeoff, killing all seven crew members.

Chamaeleon

Faint constellation of the south polar region, represented as a chameleon.

Chandrasekhar limit, or Chandrasekhar mass

In astrophysics, the maximum possible mass of a white-dwarf star. The limit depends slightly on the composition of the star but is equivalent to 1.4 times the mass of the sun. A white dwarf heavier than the Chandrasekhar limit would collapse under its own weight to form a neutron star or a black hole. The limit is named for Indian-born U.S. astrophysicist Subrahmanyan Chandrasekhar, who developed the theory of white dwarfs in the 1930s.

charge-coupled device (CCD)

Device for forming images electronically, using a layer of silicon that releases electrons when struck by incoming light. The electrons are stored in pixels and read off into a computer at the end of the exposure. CCDs have almost entirely replaced photographic film for applications, such as astrophotography, in which extreme sensitivity to light is paramount.

Chiron

Unusual solar-system object orbiting between Saturn and Uranus, discovered in 1977 by U.S. astronomer Charles T. Kowal. Initially classified as an asteroid, it is now believed to be a giant cometary nucleus, at least 200 km/120 mi across, composed of ice with a dark crust of carbon dust. It has a fifty-one-year orbit and a coma (cloud of gas and dust) caused by evaporation from its surface, resembling that of a comet. It is classified as a centaur.

chondrite

Type of meteorite characterized by chondrules, small spheres, about 1 mm/0.04 in in diameter, made up of the silicate minerals olivine and orthopyroxene.

chondrule

A small, round mass of silicate material found in chondrites (stony meteorites). Chondrules are thought to be mineral grains that condensed from hot gas in the early solar system; most were later incorporated into larger bodies from which the planets formed.

chromosphere

From the Greek words for "color" and "sphere." A layer of mostly hydrogen gas about 10,000 km/6,000 mi deep above the visible surface of the sun (the photosphere). It appears pinkish red during eclipses of the sun.

Circinus

Small constellation of the Southern Hemisphere, represented as a pair of compasses.

circumpolar

A description applied to celestial objects that remain above the horizon at all times and do not set as seen from a given location. The amount of sky that is circumpolar depends on the observer's latitude on Earth. At Earth's poles, all of the visible sky is circumpolar; at Earth's equator, none of it is.

circumpolar star

Any star that appears to circle around the Earth's north and/or south poles without either rising or setting, as a result of the motion of Earth. The radius of the area around the north pole in which such stars occur is equal to the latitude of the place of observation. Thus, at the pole, all visible stars are circumpolar, while, at the equator, none are. From London, for example, all stars within 52° of the north pole are circumpolar.

Clarke orbit

Alternative name for geostationary orbit, an orbit 35,900 km/22,300 mi high, in which satellites circle at the same speed as Earth turns. This orbit was first suggested by English space writer Arthur C. Clarke in 1945.

CNO cycle

See carbon cycle.

cold-dark-matter theory

In cosmology, the notion that the bulk of the matter in the universe is in the form of dark, unseen material consisting of slow-moving particles. The gravitational clumping of this dark matter in the early universe is what may have led to the formation of clusters and super clusters of galaxies.

color index

In astronomy, a measure of the color of a star made by comparing its brightness through different colored filters. It is defined as the difference between the magnitude of the star measured through two standard photometric filters. Color index is directly related to the surface temperature of a star and its spectral classification.

Columba

Small constellation of the Southern Hemisphere, represented as a dove. Its earlier name, Columba Noachii, identifies it with the dove sent forth by Noah after the Flood.

coma

In astronomy, the hazy cloud of gas and dust that surrounds the nucleus of a comet.

Coma Berenices

Constellation of the Northern Hemisphere, represented as Queen Berenice's hair. It was named in the third century B.C. to appease Queen Berenice after her hair, which she had sacrificed to Aphrodite, the Greek goddess of love and beauty, had been stolen from the temple. Many of the brighter stars in the constellation belong to an extended open cluster, but the term "Coma cluster" usually refers to a concentration of the galaxies that abound in this part of the sky.

comet

Small, icy body orbiting the sun, usually on a highly elliptical path. A comet consists of a central nucleus a few miles across and has been likened to a dirty snowball because it consists mostly of ice mixed with dust. As the comet approaches the sun, the nucleus heats up, releasing gas and dust that form a tenuous coma, up to 100,000 km/60,000 mi wide, around the nucleus. Gas and dust stream away from the coma to form one or more tails, which may extend for millions of miles.

Comets are believed to have been formed at the birth of the solar system. Billions of them may reside in a halo (the Oort cloud) beyond Pluto. The gravitational effect of passing stars pushes some toward the sun, when they eventually become visible from Earth. Most comets swing around the sun and return to distant space, never to be seen again for thousands or millions of years, although some, called periodic comets, have their orbits altered by the gravitational pull of the planets so that they reappear every two hundred years or less. Of the eight hundred or so comets whose orbits have been calculated, about 160 are periodic. The brightest is Halley's comet. The one with the shortest-known period is Encke's comet, which orbits the sun every 3.3 years. A dozen or more comets are discovered every year, some by amateur astronomers.

A vast amount of data concerning comets and their orbits have been accumulated, much of it consistent with the hypothesis that comets have their origin in a reservoir of appropriate material on the confines of the solar system. This material and the lumps into which it accretes move round the sun in long-period orbits, relatively few of which have perihelion distances (shortest distance to the sun) of less than 50 astronomical units. Every now and again, however, some of these orbits are perturbed, either by mutual action or by the action of passing stars, so that lumps traveling in them will pass sufficiently close to the sun to become visible as comets.

Comets are divided into two classes, according to their orbital period: Those with periods less than two hundred years are known as short-period, while the remainder are long-period comets. The orbits of the long-period comets are inclined at all angles to the ecliptic, and about equal numbers of them are direct and retrograde. The short-period comets, on the other hand, move mainly in direct orbits that lie close to the mean plane of the solar system and have aphelion distances (furthest ditance from the sun) close to the orbit of Jupiter. Such comets are sometimes referred to as belonging to Jupiter's family, and there are similar, but less numerous, families associated with the other major planets. There is little doubt that they are long-period comets that have been captured. B. G. Marsden's *Catalog of Cometary Orbits* (1972) lists 97 short-period comets and 503 long-period ones.

Most comets are named for their discoverers and denoted by letters in order of discovery each year, but they are subsequently numbered in order of their perihelion passage (the point at which they are closest to the sun). For example, the Arend-Roland comet was 1956h but became 1957 III. Some of the orbits of the long-period comets so closely resemble each other that there is little doubt that they are traced out by fragments of a former bigger comet that broke into pieces as it passed round the sun. One of the best known of these groups is that of the bright, sun-grazing comets that includes the Great Comet of 1668, 1843 I, 1880 I, 1882 II, 1887 I, 1945 VII, 1963 V, 1965 VIII, 1970 VI, and possibly one or two others. These comets passed through the solar corona and, near perihelion, were bright enough to be observed in daylight.

Few of the short-period comets show conspicuous tails, presumably because they have lost most of their volatile material. Some disintegrate and are not seen again, though some of the fragments may cause periodic meteor showers.

Comet Hale-Bopp (C/1995 01)

Large and exceptionally active comet, which in March 1997 made its closest flyby to Earth since 2000 B.C., coming within 190 million km/118 million mi. It has an icy nucleus of approximately 40 km/25 mi and an extensive gas coma (when close to the sun, Hale-Bopp released 10 metric tons/11 tons of gas every second). Unusually, Hale-Bopp has three tails: one consisting of dust particles, one of charged particles, and a third of sodium particles. Comet Hale-Bopp was discovered independently in July 1995 by two amateur U.S. astronomers, Alan Hale and Thomas Bopp.

Comet Shoemaker-Levy 9

A comet that crashed into Jupiter in July 1994. The fragments crashed into Jupiter at 60 kps/37 mps over the period July 16–22, 1994. The impacts occurred on the far side of Jupiter, but the impact sites came into view of Earth about twenty-five minutes later. Analysis of the impacts shows that most of the pieces were solid bodies about 1 km/0.6 mi in diameter but that at least three of them were clusters of smaller objects.

When first sighted on March 24, 1993, by U.S. astronomers Carolyn and Eugene Shoemaker and David Levy, it was found to consist of at least twenty-one fragments in an unstable orbit around Jupiter. It is believed to have been captured by Jupiter about 1930 and fragmented by tidal forces on passing within 21,000 km/13,050 mi of the planet in July 1992.

communications satellite

Relay station in space for sending telephone, television, telex, and other messages around the world. Messages are sent to and from the satellites via ground stations. Most communications satellites are in geostationary orbit, appearing to hang fixed over one point on Earth's surface.

The first satellite to carry TV signals across the Atlantic Ocean was *Telstar,* which was launched by the United States on July 10, 1962. The world is now linked by a system of communications satellites called *Intelsat*. Other satellites are used by individual countries for internal communications or for business or military use. A new generation of satellites, called direct-broadcast satellites, are powerful enough to transmit direct to small domestic aerials. The power for such satellites is produced by solar cells. The total energy requirement of a satellite is small; a typical communications satellite needs about 2 kW of power, the same as an electric heater.

conjunction
In astronomy, the alignment of two celestial bodies as seen from Earth. A superior planet (or other object) is in conjunction when it lies behind the sun. An inferior planet (or other object) comes to inferior conjunction when it passes between Earth and the sun; it is at superior conjunction when it passes behind the sun. Planetary conjunction takes place when a planet is closely aligned with another celestial object, such as the moon, a star, or another planet. Because the orbital planes of the inferior planets are tilted with respect to that of Earth, they usually pass either above or below the sun at inferior conjunction. If they line up exactly, a transit will occur.

constellation
One of the eighty-eight areas into which the sky is divided for the purposes of identifying and naming celestial objects. The first constellations were simple, arbitrary patterns of stars in which early civilizations visualized gods, sacred beasts, and mythical heroes. The constellations in use today are derived from a list of forty-eight known to the ancient Greeks, who inherited some from the Babylonians. The current list of eighty-eight constellations was adopted by the International Astronomical Union, astronomy's governing body, in 1930.

Cor Caroli, or Alpha Canum Venaticorum
The brightest star in Canes Venatici. It was named in honor of the English king Charles II, as it was supposed to have been especially bright on the eve of his return to London from The Hague in the Netherlands in May 1660.

corona
Faint halo of hot (about 2 million°C/3.6 million°F) and tenuous gas around the sun, which boils from the surface. It is visible at solar eclipses or through a coronagraph, an instrument that blocks light from the sun's brilliant disk. Gas flows away from the corona to form the solar wind.

Corona Australis
Small constellation of the Southern Hemisphere, popularly known as Southern Crown, located near the constellation Sagittarius. It is similar in size and shape to Corona Borealis but is not as bright.

Corona Borealis
Small but easily recognizable constellation of the Northern Hemisphere, popularly known as Northern Crown, between Hercules and Boötes, traditionally

identified with the jeweled crown of Ariadne that was cast into the sky by Bacchus in Greek mythology. Its brightest star is Alphecca (or Gemma), which is 78 light-years from Earth. It contains several variable stars. R. Coronae Borealis is normally fairly constant in brightness but fades at irregular intervals and stays faint for a variable length of time. T. Coronae Borealis is normally faint but very occasionally blazes up and, for a few days, may be visible to the naked eye. It is a recurrent nova.

Corvus
Small constellation of the Southern Hemisphere, represented as a crow. Its four brightest stars are easily recognizable. It is usually seen as being perched on Hydra.

cosmic background radiation, or 3° radiation
Electromagnetic radiation left over from the original formation of the universe in the Big Bang between ten billion and twenty billion years ago. It corresponds to an overall background temperature of 2.73K (–270.4°C/–454.7°F), or 3°C above absolute zero. In 1992, the U.S. *Cosmic Background Explorer* (COBE) satellite detected slight "ripples" in the strength of the background radiation that are believed to mark the first stage in the formation of galaxies. Cosmic background radiation was first detected in 1965 by German-born U.S. radio engineer Arno Penzias and U.S. radio astronomer Robert Wilson, who in 1978 shared the Nobel Prize for Physics for their discovery.

cosmic radiation
Streams of high-energy particles from outer space, consisting of protons, alpha particles, and light nuclei, which collide with atomic nuclei in Earth's atmosphere and produce secondary nuclear particles (chiefly mesons, such as pions and muons) that shower Earth. Those of lower energy seem to be galactic in origin; those of high energy, of extragalactic origin. The galactic particles may come from supernova explosions or pulsars. At higher energies, other sources are necessary, possibly the giant jets of gas that are emitted from some galaxies.

cosmogony
From the Greek for "universe" and "creation"; the study of the origin and evolution of cosmic objects, especially the solar system.

cosmological principle
A hypothesis that states that any observer anywhere in the universe has the same view that we have—that is, that the universe is not expanding from any center but that all galaxies are moving away from one another.

cosmology
Branch of astronomy that deals with the structure and evolution of the universe as an ordered whole. Its method is to construct "model universes" mathematically and compare their large-scale properties with those of the observed uni-

verse. Modern cosmology began in the 1920s with the discovery that the universe is expanding, which suggested that it began in an explosion, the Big Bang. An alternative—now discarded—view, the steady-state theory, claimed that the universe has no origin but is expanding because new matter is being continually created.

There are a number of differences in the conclusions that can be drawn from the steady-state and the Big Bang theories. For example, the number of galaxies per unit volume should not change with distance if the steady-state theory is correct, but should increase with distance if an evolutionary theory is correct, since, according to the latter, in looking over a distance we are also looking back in time with the universe gradually getting more compact. The latest counts of faint radio sources do seem to indicate an increase in the number per unit volume with distance, which supports an evolutionary model. Another different conclusion: According to the steady-state theory, the mixture of old and new galaxies should be the same at both small and large distances, while evolutionary theories hold that the proportion of young objects should increase with distance. Although the nature of quasars is not yet understood, they do appear to be young objects, and they are found only with large red shifts (meaning far back in time), so their very existence seems to disprove the steady-state theory.

Another piece of evidence for the Big Bang theory is the cosmic background radiation, which was first observed in 1965 and can be interpreted as radiation predicted as a necessary consequence of the Big Bang. If this identification is accepted, the present temperature of the background radiation can be used to calculate what that of the primeval atom must have been in the early stages of its expansion and, thus, predict what the initial ratio of hydrogen to helium should have been, a prediction that turns out to be quite consistent with observation.

Finally, as far as present observational data are concerned, there is no indication that the rate of expansion of the universe is in any way slowing down and, thus, no evidence to support the oscillating models of the universe that had sometimes been suggested when it was conceived as possible that the rate of expansion could not only be slowed down but actually reversed.

cosmonaut
Term used in the West for any astronaut from the former Soviet Union.

Cosmos
Name used from the early 1960s for nearly all Soviet artificial satellites. More than twenty-three hundred *Cosmos* satellites had been launched by mid-1995.

Crab Nebula
Cloud of gas 6,000 light-years from Earth, in the constellation Taurus. It is the remains of a star that, according to Chinese records, exploded as a supernova observed as a brilliant point of light on July 4, 1054. At its center is a pulsar that flashes thirty times a second. The nebula was named for its crablike shape.

The Crab Nebula is a powerful radio and X-ray source. Optically, it appears as a diffuse elliptical area on which is superimposed an intricate network of bright filaments. Observations show that it is increasing in size; its present dimensions are of the order of 10 light-years. The light is highly polarized, suggesting the presence of strong magnetic fields. This suggestion is strengthened by the fact that the diffuse portion is emitting radiation throughout the whole electromagnetic spectrum from radio to gamma waves, the energy coming from the pulsar near the center.

Crater
Small constellation of the Southern Hemisphere, represented as a cup; it is associated in mythology with Hydra and Corvus.

crater
Bowl-shaped depression in the ground, usually round and with steep sides. Craters are formed by explosive events, such as the eruption of a volcano, the explosion of a bomb, or the impact of a meteorite. Earth's moon has more than 300,000 craters that measure more than 1 km/6 mi in diameter, formed by meteorite bombardment; similar craters on Earth have mostly been worn away by erosion. Craters are found on many other bodies in the solar system.

Studies at the Jet Propulsion Laboratory in Pasadena, California, have shown that craters produced by impact or by volcanic activity have distinctive shapes, enabling astronomers to distinguish likely methods of crater formation on planets in the solar system. Unlike volcanic craters, impact craters have a raised rim and a central peak and are almost always circular, regardless of the meteorite's angle of incidence.

crescent
Curved shape of the moon when it appears less than half illuminated. It also refers to any object or symbol resembling the crescent moon.

critical density
In cosmology, the minimum average density that the universe must have for it to stop expanding at some point in the future. The precise value depends on Hubble's constant and so is not fixed, but it is approximately between 10^{-29} and 2×10^{29} g/cm^3, equivalent to a few hydrogen atoms per cubic meter. The density parameter (symbol $\grave{U}$) is the ratio of the actual density to the critical density. If $\grave{U}$ is less than 1, the universe is open and will expand forever. If $\grave{U}$ is greater than 1, the universe is closed, and the expansion will eventually halt, to be followed by a contraction. Current estimates from visible matter in the universe indicate that $\grave{U}$ is about 0.01, well below critical density, but unseen dark matter may be sufficient to raise $\grave{U}$ to somewhere between 0.1 and 2.

Crux
Constellation of the Southern Hemisphere, popularly known as the Southern Cross; it is the smallest of the eighty-eight constellations but one of the brightest. It is also one of the best known, since it is represented on the flags of Australia and

New Zealand. Its brightest stars are Alpha Crucis (or Acrux), a double star about 360 light-years from Earth, and Beta Crucis (or Mimosa) which is about 420 light-years from Earth. Near Beta Crucis lies a glittering star cluster known as the Jewel Box. The constellation also contains the Coalsack, a dark nebula silhouetted against the bright starry background of the Milky Way.

Crux, situated in one of the brightest sections of the Milky Way, originally formed part of Centaurus; it was not named as a distinct asterism until the fifteenth century and not regarded as a separate constellation until the seventeenth.

Cygnus
Large prominent constellation of the Northern Hemisphere, represented as a swan. Its brightest star is first-magnitude Alpha Cygni (or Deneb). Beta Cygni (or Albireo) is a yellow and blue double star, visible through small telescopes. The constellation contains the North America Nebula (named for its shape), the Veil Nebula (the remains of a supernova that exploded about fifty thousand years ago), Cygnus A (apparently a double galaxy, a powerful radio source, and the first radio star to be discovered), and the X-ray source Cygnus X-1, thought to mark the position of a black hole. The area is rich in high-luminosity objects, nebulae, and clouds of obscuring matter. Deneb marks the tail of the swan, which is depicted as flying along the Milky Way. Some of the brighter stars form the Northern Cross, the upright being defined by Alpha, Gamma, Eta, and Beta, and the crosspiece by Delta, Gamma, and Epsilon Cygni.

dark cloud
In astronomy, a cloud of cold dust and gas seen in silhouette against background stars or an HII region.

dark matter
Matter that, according to late-twentieth-century theories of cosmology, makes up 90–99 percent of the mass of the universe but so far remains undetected. Dark matter, if shown to exist, would explain many so far unexplained gravitational effects in the movement of galaxies. Theories of the composition of dark matter include unknown atomic particles (cold dark matter), or fast-moving neutrinos (hot dark matter), or a combination of both.

In 1993, astronomers identified part of the dark matter in the form of stray planets and brown dwarfs and, possibly, stars that have failed to light up. These objects are known as MACHOs (massive astrophysical compact halo objects) and, according to U.S. astronomers in 1996, make up approximately half of the dark matter in the Milky Way's halo.

David Dunlap Observatory
Canadian observatory at Richmond Hill, Ontario, operated by the University of Toronto, with a 1.88-m/74-in reflector, the largest optical telescope in Canada; it opened in 1935.

day

Time taken for Earth to rotate once on its axis. The solar day is the time that Earth takes to rotate once relative to the sun. It is divided into twenty-four hours and is the basis of our civil day. The sidereal day is the time that Earth takes to rotate once relative to the stars. It is 23.93 hours, just less than four minutes shorter than the solar day, because the sun's position against the background of stars as seen from Earth changes as Earth orbits it.

declination

In astronomy, the coordinate on the celestial sphere (imaginary sphere surrounding Earth) that corresponds to latitude on Earth's surface. Declination runs from 0° at the celestial equator to 90° at the north and south celestial poles.

Deimos

One of the two moons of Mars. It is irregularly shaped, $15 \times 12 \times 11$ km/$9 \times 7.5 \times 7$ mi, orbits at a height of 24,000 km/15,000 mi every 1.26 days, and is not as heavily cratered as the other moon, Phobos. Deimos was discovered in 1877 by U.S. astronomer Asaph Hall and is thought to be an asteroid captured by Mars's gravity.

Delphinus

Small but fairly conspicuous constellation on the celestial equator, represented as a dolphin.

Delta rocket

U.S. rocket used to launch many scientific and communications satellites since 1960, based on the *Thor* ballistic missile. Several increasingly powerful versions produced as satellites became larger and heavier. Solid-fuel boosters were attached to the first stage to increase lifting power.

Deneb, or Alpha Cygni

Brightest star in the constellation Cygnus and the nineteenth-brightest star in the night sky. It is one of the greatest supergiant stars known, with a true luminosity of about sixty thousand times that of the sun. Deneb is about 1,800 light-years from Earth. The name Deneb is derived from the Arabic word for "tail."

density wave

In astrophysics, a concept proposed to account for the existence of spiral arms in galaxies. In the density-wave theory, stars in a spiral galaxy move in elliptical orbits in such a way that they crowd together in waves of temporarily enhanced density that appear as spiral arms. The idea was first proposed by Swedish astronomer Bertil Lindblad in the 1920s and developed by U.S. astronomers C. C. Lin and Frank Shu in the 1960s.

direct motion

See prograde, or direct motion.

disk

In astronomy, the flat, roughly circular region of a spiral or a lenticular (lens shaped) galaxy containing stars, nebulae, and dust clouds orbiting about the nucleus. Disks contain predominantly young stars and regions of star formation. The disk of our own galaxy is seen from Earth as the band of the Milky Way.

distance modulus

A method of finding the distance to an object in the universe, such as a star or a galaxy, using the difference between the actual and the apparent brightness of the object. The actual brightness is deduced from the object's type and its size. The apparent brightness is obtained by direct observation.

Dog Star

See Sirius, or Dog Star, or Alpha Canis Majoris.

Dominion Astrophysical Observatory

Canadian observatory near Victoria, British Columbia, the site of a 1.85-m/73-in reflector, opened in 1918, operated by the National Research Council of Canada. The associated Dominion Radio Astrophysical Observatory at Penticton, British Columbia, opened in 1959, operates a 26-m/85-ft radio dish and an aperture-synthesis radio telescope.

Dorado

Constellation of the Southern Hemisphere, represented as a goldfish or a sword-fish. It is easy to locate, since the Large Magellanic Cloud marks its southern border. Its brightest star is Alpha Doradus, just under 200 light-years from Earth. One of the most conspicuous objects in the Large Magellanic Cloud is the Great Looped Nebula that surrounds 30 Doradus.

double star

Two stars that appear close together. Many stars that appear single to the naked eye appear double when viewed through a telescope. Some double stars attract each other due to gravity and orbit each other, forming a genuine binary star, but other double stars are at different distances from Earth and lie in the same line of sight only by chance. Through a telescope, both types look the same.

Double stars of the second kind, which are of little astronomical interest, are referred to as optical pairs; those of the first as physical pairs or, more usually, visual binaries. They are the principal source from which our knowledge of stellar masses is derived.

Draco

A large but faint constellation, represented as a dragon coiled around the north celestial pole. Due to precession (the Earth's axial wobble), the star Alpha Draconis (or Thuban) was the pole star 4,800 years ago. This star seems to have faded, for it is no longer the brightest star in the constellation, as it was at the beginning of the seventeenth century. Gamma Draconis is more than a magnitude brighter. It was extensively observed in the early eighteenth century by English astronomer James Bradley, who, from its apparent changes in position, discovered the aberration of starlight and nutation.

Earth

Third planet from the sun. It is almost spherical, flattened slightly at the poles, and is composed of three concentric layers: the core, the mantle, and the crust. About 70 percent of the surface (including the north and south polar icecaps) is covered with water. Earth is surrounded by a life-supporting atmosphere and is the only planet on which life is known to exist.

mean distance from the sun:
149.5 million km/92.9 million mi

equatorial diameter:
12,756 km/7,923 mi

circumference:
40,070 km/24,900 mi

rotation period:
23.93 hours

year:
(Complete orbit, or sidereal period) 365.26 days. Earth's average speed around the sun is 30 kps/18.5 mps; the plane of its orbit is inclined to its equatorial plane at an angle of 23.5°, the reason for the changing seasons.

atmosphere:
Nitrogen (78.09 percent), oxygen (20.95 percent), argon (0.93 percent), carbon dioxide (0.03 percent), and less than 0.0001 percent neon, helium, krypton, hydrogen, xenon, ozone, and radon

surface:
Land surface 150 million km^2/57.5 million mi^2 (greatest height above sea level 8,872-m/29,118-ft Mount Everest); water surface 361 million km^2/139.4 million mi^2 (greatest depth 11,034-m/36,201-ft Mariana Trench in the Pacific Ocean). The interior is thought to consist of an inner core, about 2,600 km/1,600 mi in diameter, of solid iron and nickel; an outer core, about 2,250 km/1,400 mi thick, of molten iron and nickel; and a mantle, about 2,900 km/1,800 mi thick, of mostly solid rock separated from Earth's crust by the Mohorovicic discontinuity (the boundary that separates the Earth's crust and mantle). The crust and the topmost layer of the mantle form twelve major moving plates, some of which carry the continents. The plates are in constant, slow motion, called tectonic drift. U.S. geo-

physicists announced in 1996 that they had detected a difference in the spinning time of Earth's core and the rest of the planet; the core is spinning slightly faster.

satellite:
The moon

age:
4.6 billion years. Earth was formed with the rest of the solar system by consolidation of interstellar dust. Life began 3.5–4 billion years ago.

eclipse
Passage of an astronomical body through the shadow of another. The term is usually employed for solar and lunar eclipses, which may be either partial or total, but also, for example, for eclipses by Jupiter of its satellites. An eclipse of a star by a body in the solar system is called an occultation.

A solar eclipse occurs when the moon passes in front of the sun as seen from Earth and can happen only at new moon. During a total eclipse, which can last up to 7.5 minutes, the sun's corona can be seen. When the moon is at its farthest from Earth, it does not completely cover the face of the sun, leaving a ring of sunlight visible. This is an annular eclipse (from the Latin word *annulus,* "ring"). Between two and five solar eclipses occur each year.

A lunar eclipse occurs when the moon passes into the shadow of Earth, becoming dim until emerging from the shadow. Lunar eclipses may be partial or total, and they can happen only at full moon. Total lunar eclipses last for up to 1.7 hours; the maximum number each year is three.

ecliptic
Path, against the background of stars, that the sun appears to follow each year as it is orbited by Earth. It can be thought of as the plane of Earth's orbit projected onto the celestial sphere (imaginary sphere around Earth). The ecliptic is tilted at 23.5° with respect to the celestial equator, a result of the tilt of Earth's axis relative to the plane of its orbit around the sun.

ecliptic coordinates
A system for measuring the position of astronomical objects on the celestial sphere with reference to the plane of Earth's orbit, the ecliptic.

Ecliptic latitude (symbol ß) is measured in degrees from the ecliptic (ß = 0°) to the north (ß = 90°) and south (ß = – 90°) ecliptic poles.

Ecliptic longitude (symbol ë) is measured in degrees eastward along the ecliptic (ë = 0° to 360°) from a fixed point known as the first point of Aries, or the vernal equinox. Ecliptic coordinates are often used to measure the positions of the sun and the planets with respect to Earth.

Ecliptic latitude and longitude are sometimes known as celestial latitude and longitude. The ecliptic longitude of the sun (solar longitude) is a convenient measure of the position of Earth in its orbit.

Edwards Air Force Base
United States Air Force center in California, situated on a dry lake bed, often used as a landing site by the U.S. space shuttle.

Effelsberg Radio Observatory
Site, near Bonn, Germany, of the world's largest fully steerable radio telescope, the 100-m/328-ft radio dish of the Max Planck Institute for Radio Astronomy, opened in 1971.

ejecta
Any material thrown out of a crater due to volcanic eruption or the impact of a meteorite or other object. Ejecta from impact craters on the moon often form long bright streaks known as rays, which, in some cases, can be traced for thousands of kilometers across the lunar surface.

elliptical galaxy
One of the main classes of galaxy in the Hubble classification, characterized by a featureless elliptical profile. Unlike spiral galaxies, elliptical galaxies have very little gas or dust, and no stars have recently formed within them. They range greatly in size from giant ellipticals, which are often found at the centers of clusters of galaxies and may be strong radio sources, to tiny dwarf ellipticals, containing about a million stars, which are the most common galaxies of any type. More than 60 percent of known galaxies are elliptical.

elongation
The angular distance between the sun and a planet or other solar-system object. This angle is 0° at conjunction, 90° at quadrature, and 180° at opposition.

emission line
Bright line in the spectrum of a luminous object caused by atoms emitting light at sharply defined wavelengths.

Encke's comet
Comet with the shortest-known orbital period, 3.3 years. It is named for German mathematician and astronomer Johann Franz Encke, who calculated its orbit in 1819 from earlier sightings. It was first seen in 1786 by the French astronomer Pierre Méchain. It is the parent body of the Taurid meteor shower, and a fragment of it may have hit Earth in the 1908 Tunguska Event. In 1913, it became the first comet to be observed throughout its entire orbit when it was photographed near aphelion (the point in its orbit farthest from the sun) by astronomers at Mount Wilson Observatory, near Los Angeles, California.

Energiya
The most powerful Soviet space rocket, first launched on May 15, 1987. Used to launch the Soviet space shuttle, the *Energiya* booster is capable, with the use of

strap-on boosters, of launching payloads of up to 190 metric tons/210 tons into Earth orbit.

Epsilon Aurigae
Eclipsing binary star in the constellation Auriga. One of the pair is an "ordinary" star, but the other seems to be an enormous distended object whose exact nature remains unknown. The period (time between eclipses) is twenty-seven years, the longest of its kind. The last eclipse was 1982–1984.

equatorial coordinates
A system for measuring the position of astronomical objects on the celestial sphere with reference to the plane of Earth's equator. Declination (symbol Ä), analogous to latitude, is measured in degrees from the equator to the north (Ä = 90°) or south (Ä = –90°) celestial poles. Right ascension (symbol á), analogous to longitude, is normally measured in hours of time (á = 0 h to 24 h) eastward along the equator from a fixed point known as the first point of Aries, or the vernal equinox.

equatorial mounting
A method of mounting a telescope to simplify the tracking of celestial objects. One axis (the polar axis) is mounted parallel to the rotation axis of Earth so that the telescope can be turned about it to follow objects across the sky. The declination axis moves the telescope in declination and is clamped before tracking begins. Another advantage over the simpler altazimuth mounting is that the orientation of the image is fixed, permitting long-exposure photography.

equinox
The points in spring and fall at which the sun's path, the ecliptic, crosses the celestial equator, so that day and night are of approximately equal length. The vernal equinox occurs about March 21; the autumnal equinox, about September 23.

Equuleus
Small, faint constellation on the celestial equator, represented as a foal.

Eridanus
The sixth-largest constellation, which meanders from the celestial equator deep into the Southern Hemisphere of the sky. Eridanus is represented as a river. Its brightest star is Achernar.

Eros
An asteroid, discovered in 1898 by G. Witt, that can pass 22 million km/14 million mi from Earth, as observed in 1975. Eros was the first asteroid to be discovered that has an orbit coming within that of Mars. It is elongated, measures about 36 × 12 km/22 × 7 mi, rotates around its shortest axis every 5.3 hours, and orbits the sun every 1.8 years. The *Near Earth Asteroid Rendezvous* (NEAR), launched

in February 1996 by the U.S. National Aeronautics and Space Administration (NASA), was scheduled to reach Eros in February 1999 and spend a year circling the asteroid in an attempt to determine what it is made of.

ESA
See European Space Agency (ESA).

escape velocity
In physics, the minimum velocity with which an object must be projected for it to escape from the gravitational pull of a planetary body. In the case of Earth, the escape velocity is 11.2 kps/6.9 mps; the moon, 2.4 kps/1.5 mps; Mars, 5 kps/3.1 mps; and Jupiter, 59.6 kps/37 mps.

Eta Tauri
See Alcyone, or Eta Tauri.

Europa
The fourth-largest moon of the planet Jupiter, 3,140 km/1,950 mi in diameter, orbiting 671,000 km/417,000 mi from the planet every 3.55 Earth days. It is covered by ice and criss-crossed by thousands of thin cracks, each around 50,000 km/30,000 mi long.

The robot probe *Galileo,* launched by the United States in 1989, began circling Europa in February 1997 and was expected to send back around eight hundred images from fifty different sites by 1999. One of the first discoveries was that what were thought to be cracks covering the surface of the moon are, in fact, low ridges. Further investigation is needed to determine their origin. Analysis in 1998 of high-resolution images from *Galileo* suggests that Europa's icy crust may hide a vast ocean warm enough to support life. The National Aeronautics and Space Administration (NASA) has announced plans to launch the *Europa Observer* in 2003 to search for such water.

European Southern Observatory
Observatory operated jointly by Belgium, Denmark, France, Germany, Italy, the Netherlands, Sweden, and Switzerland, with headquarters near Munich, Germany. Its telescopes, located at La Silla, Chile, include a 3.6-m/142-in reflector, opened in 1976, and the 3.5-m/138-in New Technology Telescope, opened in 1989. By 1988, work began on the Very Large Telescope (VLT), at Cerro Paranal, Chile, consisting of four 8-m/26-ft reflectors mounted independently but capable of working in combination. The "first light" of the first 8-m/26-ft telescope occurred in May 1998. The VLT will not be fully operational until 2001.

European Space Agency (ESA)
Organization of European countries (Austria, Belgium, Britain, Denmark, Finland, France, Germany, Ireland, Italy, the Netherlands, Norway, Spain, Sweden, and Switzerland) that engages in space research and technology. It was founded

in 1975, with headquarters in Paris, France. ESA has developed various scientific and communications satellites, the *Giotto* space probe, and the *Ariane* rockets. It built *Spacelab* and plans to build its own space station, *Columbus,* for attachment to a U.S. space station. ESA's Earth-sensing satellite *ERS-2* was launched successfully in 1995. It works in tandem with *ERS-1,* which was launched in 1991, and has improved measurements of global ozone.

exobiology
Study of life-forms that may possibly exist elsewhere in the universe and of the effects of extraterrestrial environments on Earth organisms. Techniques include space-probe experiments designed to detect organic molecules and the monitoring of radio waves from other star systems.

Explorer
Series of U.S. scientific satellites. *Explorer 1,* launched in January 1958, was the first U.S. satellite in orbit and discovered the Van Allen radiation belts around Earth.

filament
In astronomy, a dark, winding feature occasionally seen on images of the sun in hydrogen light. Filaments are clouds of relatively cool gas suspended above the sun by magnetic fields and seen in silhouette against the hotter photosphere below. During total eclipses, they can be seen as bright features against the sky at the edge of the sun, where they are known prominences.

fireball
In astronomy, a very bright meteor, often bright enough to be seen in daylight and occasionally leading to the fall of a meteorite. Some fireballs are caused by satellites or other space debris burning up in Earth's atmosphere.

fixed star
Description sometimes applied to a star, especially in emphasizing its difference from a planet ("wandering star"). Though most stars are in rapid motion through space, their great distance from Earth ensures that their relative configurations do not change appreciably over a thousand years.

focus
In astronomy, either of two points lying on the major axis of an elliptical orbit on either side of the center. One focus marks the center of mass of the system, and the other is empty. In a circular orbit, the two foci coincide at the center of the circle; in a parabolic orbit, the second focus lies at infinity.

Fomalhaut, or Alpha Piscis Austrini
The brightest star in the southern constellation Piscis Austrinus and the eighteenth-brightest star in the night sky. It is 22 light-years from Earth, with a true

luminosity thirteen times that of the sun. Fomalhaut is one of a number of stars around which the *Infrared Astronomy Satellite* (IRAS) detected excess infrared radiation, presumably from a region of solid particles around the star. This material may be a planetary system in the process of formation.

footprint
Of a satellite, the area of Earth over which its signals can be received.

forbidden line
Emission line seen in the spectra of certain astronomical objects that is not seen under the conditions prevailing in laboratory experiments. Forbidden lines indicate that the hot gas emitting them is at extremely low density. They are seen, for example, in the tenuous gas of the sun's corona, in HII regions (regions of hot ionized hydrogen), and in the nucleuses of certain active galaxies.

Fornax
Inconspicuous constellation of the Southern Hemisphere, represented as a furnace. It was named by French astronomer Nicolas Lacaille in honor of the French chemist Antoine Lavoisier.

galactic cluster
See open cluster, or galactic cluster.

galactic coordinates
A system for measuring the position of astronomical objects on the celestial sphere with reference to the galactic equator (or great circle).

Galactic latitude (symbol b) is measured in degrees from the galactic equator (b = 0°) to the north (b = 90°) and south (b = – 90°) galactic poles.

Galactic longitude (symbol l) is measured in degrees eastward (l = 0° to 360°) from a fixed point in the constellation of Sagittarius that approximates to the center of the galaxy. Galactic coordinates are often used when astronomers are studying the distribution of material in the galaxy.

galactic halo
The outer, sparsely populated region of a galaxy, roughly spheroid in shape and extending far beyond the bulk of the visible stars. In our own galaxy, the halo contains the globular clusters and may harbor large quantities of dark matter.

galactic plane
A plane passing through the sun and the center of the galaxy defining the midplane of the galactic disk. Viewed from Earth, the galactic plane is a great circle (galactic equator) marking the approximate center line of the Milky Way.

galaxy
Congregation of millions or billions of stars, held together by gravity.

spiral galaxies
Spiral galaxies, such as the Milky Way, are flattened in shape, with a central bulge of old stars surrounded by a disk of younger stars, arranged in spiral arms like a Catherine wheel.

barred spirals
Barred spirals are spiral galaxies that have a straight bar of stars across their center, from the ends of which the spiral arms emerge. The arms of spiral galaxies contain gas and dust from which new stars are still forming.

elliptical galaxies
Elliptical galaxies contain old stars and very little gas. They include the most massive galaxies known, containing a trillion stars. At least some elliptical galaxies are thought to be formed by mergers between spiral galaxies. There are also irregular galaxies. Most galaxies occur in clusters, containing anything from a few to thousands of members.

Our own galaxy, the Milky Way, is about 100,000 light-years in diameter and contains at least 100 billion stars. It is a member of a small cluster, the Local Group. The sun lies in one of its spiral arms, about 25,000 light-years from the center.

By the end of a five-year study in 1995, U.S. astronomers had identified six hundred previously uncataloged galaxies, mostly 200–400 million light-years away, leading to the conclusion that there may be 30–100 percent more galaxies than previously estimated. Two galaxies were discovered obscured by galactic dust at the edge of the Milky Way. One, named MB1, is a spiral galaxy 17,000 light-years across; the other, MB2, is an irregular-shaped dwarf galaxy about 4,000 light-years across. In 1996, U.S. astronomers discovered a further new galaxy 17 million light-years away. The galaxy, NGC2915, is a blue compact-dwarf galaxy, and 95 percent of its mass is in the form of dark matter. In 1997, an international team of astronomers detected the farthest-known object in the universe, which is a galaxy lying 13 billion light-years away.

Galaxies vary in size, structure, and luminosity and, like stars, are found alone, in pairs, or in clusters. As these systems are very remote, they appear in telescopes as hazy, nebulous objects and were first described as nebulae. Later, when their remoteness was understood, they were known as "island universes" or "extragalactic nebulae."

Only two, the Magellanic Clouds, are easily visible to the naked eye. The next brightest, the Andromeda galaxy, is just visible. About thirty-five of the brightest galaxies appear in the list compiled by the French astronomer Charles Messier, and several thousand are in the *New General Catalog*. More than 100 million can be photographed with modern telescopes.

distances from Earth
About twenty galaxies are known to be within 2.5 million light-years away from Earth, and several thousand within 50 million light-years. The distances

of those closer than 10 million light-years can be estimated from the brightness of individual Cepheid variables if such stars can be identified. Up to about 100 million light-years away, the magnitudes of supergiants, and of novae or supernovae at maximum, can be used to determine distance. Still greater distances have been estimated by comparing the apparent magnitude of a galaxy with its absolute magnitude. The greatest distances of all are found by measuring the red shift (i.e., the lengthening of the wavelength of light from an object as it moves away) and assuming the truth of the red shift–distance relation, one of the essential dogmas of modern cosmology. Once the distance is known, it becomes possible to estimate the masses of some of the nearer galaxies. It is also possible to estimate the masses of clusters of galaxies, but the masses so found have been larger than would be expected from the sum of the masses of the visible individual galaxies. This discrepancy, sometimes known as the problem of the missing mass, has not been explained.

types of galaxy

Normal galaxies were classified by the U.S. astronomer Edwin Hubble into three basic types: spiral, elliptical, and irregular. Spiral galaxies, of which our own galaxy is a typical example, consist of a nucleus, a disk containing the spiral arms, and a halo. Spiral galaxies are classified according to the appearance of their arms. Sa spirals have a large nuclear bulge and tightly coiled spiral arms, while Sc spirals have a small nucleus and arms less tightly wound. Sb spirals are intermediate between Sa and Sc. Barred spirals are classified SBa to SBc. Elliptical galaxies are something like huge globular clusters, with no spiral arms. They are divided into eight subgroups, E0–E7, the E0s appearing spherical and the E7s the most elongated. Irregular galaxies have a chaotic appearance and show no symmetry. Irregulars are very much less frequent than spirals and ellipticals; they also tend to be smaller but brighter in proportion to their mass.

clusters

Clusters of galaxies can be roughly classified as regular or irregular. Regular clusters have spherical symmetry, central concentration, and usually at least a thousand members brighter than absolute magnitude –16. One of the nearest examples is in Corona Borealis. Irregular clusters are made up of loose groups of small clusters. Unlike the regulars, which consist almost entirely of ellipticals, the irregular clusters contain all types of galaxies. They vary greatly in content and may contain more than a thousand galaxies, as in the cluster in Virgo; or only twenty or so, as in the Local Group to which our galaxy belongs. The Local Group contains about thirty members within a region 3 million light-years across: two large Sb spirals (our own and the Andromeda galaxy); one smaller Sc spiral; fourteen ellipticals, of which ten are dwarfs; and four irregulars, of which two are the Magellanic Clouds. It has been suggested that our Local Group is only a subsection of a "Local Super Cluster." This may be centered in or near the Virgo cluster and is about 100 million light-years in diameter and 25 million light-years thick.

Galileo
Spacecraft launched from the U.S. space shuttle *Atlantis* in October 1989 on a six-year journey to Jupiter. *Galileo*'s probe entered the atmosphere of Jupiter in December 1995 and radioed information back to the orbiter for fifty-seven minutes before it was destroyed by atmospheric pressure. The orbiter continued circling Jupiter until 1997. Despite technical problems, data are still being relayed to Earth, but very slowly.

gamma-ray astronomy
The study of gamma rays from space. Much of the radiation detected comes from collisions between hydrogen gas and cosmic rays in our galaxy. Some sources have been identified, including the Crab Nebula and the Vela pulsar (the most powerful gamma-ray source detected).

Gamma rays are difficult to detect and are generally studied by use of balloon-borne detectors and artificial satellites. The first gamma-ray satellites were *SAS II*, launched in 1972, and *COS B*, launched in 1975, although gamma-ray detectors were carried on the U.S. *Apollo 15* and *16* missions, in 1971 and 1972, respectively. *SAS II* failed after only a few months, but *COS B* continued working until 1982, carrying out a complete survey of the galactic disk. The *Compton Gamma Ray Observatory* was launched by U.S. space shuttle *Atlantis* in April 1991 to study the gamma-ray sky for five years. The observatory cost $617 million and, at 16.7 metric tons/18.4 tons, was the heaviest payload ever carried by space shuttle.

Ganymede
The largest moon of the planet Jupiter and the largest moon in the solar system, 5,260 km/3,270 mi in diameter (larger than the planet Mercury). It orbits Jupiter every 7.2 Earth days at a distance of 1.1 million km/700,000 mi. Its surface is a mixture of cratered and grooved terrain. Molecular oxygen was identified on Ganymede's surface in 1994.

The U.S. space probe *Galileo* detected a magnetic field around Ganymede in 1996; this suggests it may have a molten core. *Galileo* photographed Ganymede at a distance of 7,448 km/4,628 mi. The resulting images were seventeen times clearer than those taken by the U.S. *Voyager 2* in 1979 and show the surface to be extensively cratered and ridged, probably as a result of forces similar to those that create mountains on Earth. *Galileo* also detected molecules containing both carbon and nitrogen on the surface in March 1996. Their presence may indicate that Ganymede harbored life at some time.

gas giant
Any of the four large outer planets of the solar system, Jupiter, Saturn, Uranus, and Neptune, which consist largely of gas and have no solid surface.

Gemini
Prominent zodiacal constellation in the Northern Hemisphere, represented as the twins Castor and Pollux. Its brightest star is Pollux; Castor is a system of six stars. The sun passes through Gemini from late June to late July. Each December, the

Geminid meteors radiate from Gemini. In astrology, the dates for Gemini are about May 21–June 21.

Gemini 8-Meter Telescopes Project

International project to build a pair of 8-m/26-ft astronomical telescopes on high mountain sites in Hawaii and Chile. Each telescope will have a primary mirror 8-m/26-ft in diameter and 20 cm/8 in thick constructed of ultra-low-expansion glass. The shape of the mirror will be adjusted by 120 active supports to prevent it sagging as the telescope moves across the sky. A system of adaptive optics will compensate for the distorting effects of the atmosphere, producing sharp images throughout the visual and near-infrared parts of the spectrum. The two sites—at Mauna Kea, Hawaii (4,200 m/13,784 ft high), and Cerro Pachon, Chile (2,700 m/ 8,858 ft)—have excellent observing conditions and, between them, cover the whole sky. The project is funded by the United States, Britain, Canada, Chile, Argentina, Brazil, and Australia, and the telescopes are expected to become operational in 1999–2000.

Gemini Project

U.S. space program (1965–1966) in which astronauts practiced rendezvous and docking of spacecraft and working outside their spacecraft in preparation for the *Apollo* moon landings. *Gemini* spacecraft carried two astronauts and were launched by *Titan* rockets.

geostationary orbit, or synchronous orbit

Circular path 35,900 km/22,300 mi above Earth's equator on which a satellite takes twenty-four hours, moving from west to east, to complete an orbit, thus appearing to hang stationary over one place on Earth's surface. Geostationary orbits are used particularly for communications satellites and weather satellites. They were first thought of by the English space writer Arthur C. Clarke. A geosynchronous orbit lies at the same distance from Earth but is inclined to the equator.

giant star

A class of stars located at the top right of the Hertzsprung-Russell diagram characterized by great size and luminosity. Giants have exhausted their supply of hydrogen fuel and derive their energy from the fusion of helium and heavier elements. They are anywhere from ten to three hundred times bigger than the sun with thirty to a thousand times the luminosity. The cooler giants are known as red giants.

Giotto

Space probe built by the European Space Agency to study Halley's comet. Launched by an *Ariane* rocket in July 1985, *Giotto* passed within 600 km/375 mi of the comet's nucleus on March 13, 1986. On July 2, 1990, it flew 23,000 km/ 14,000 mi from Earth, which diverted its path to encounter another comet, Grigg-Skjellerup, on July 10, 1992.

globular cluster

Spherical or near-spherical star cluster containing from ten thousand to millions of stars. About 120 globular clusters are distributed in a spherical halo around our galaxy. They consist of old stars, formed early in the galaxy's history. Globular clusters are also found around other galaxies.

Goddard Space Flight Center

Installation of the U.S. National Aeronautics and Space Administration (NASA) at Greenbelt, Maryland, responsible for the operation of NASA's unmanned scientific satellites, including the Hubble Space Telescope. It is also home of the National Space Science Data Center, a repository of data collected by satellites.

Gould's belt

A band in the sky in which the brighter blue stars appear to concentrate. It is inclined at 20° to the Milky Way, which it crosses in the constellations of Vela and Lacerta, and is especially well marked in the Southern Hemisphere.

gravitational lensing

The bending of light by a gravitational field, predicted by German-born U.S. physicist Albert Einstein's general theory of relativity. The effect was first detected in 1917, when the light from stars was found to be bent as it passed the totally eclipsed sun. More remarkable is the splitting of light from distant quasars into two or more images by intervening galaxies. In 1979, the first double image of a quasar produced by gravitational lensing was discovered, and a quadruple image of another quasar was later found.

Great Bear

Popular name for the constellation Ursa Major.

Great Red Spot

Prominent oval feature, 14,000 km/8,500 mi wide and some 30,000 km/20,000 mi long, in the atmosphere of the planet Jupiter, south of the equator. It was first observed in 1664. Space probes show it to be a counterclockwise vortex of cold clouds, colored possibly by phosphorus.

Great Wall

Array of galaxies arranged almost in a perfect plane, consisting of some two thousand galaxies (about 500 million × 200 million light-years across). It was discovered in 1989 by U.S. astronomers at the Harvard College Observatory in Cambridge, Massachusetts.

Green Bank

Site in West Virginia, of the National Radio Astronomy Observatory. Its main instruments are a 43-m/140-ft fully steerable dish, opened in 1965, and three 26-m/85-ft dishes. A 90-m/300-ft partly steerable dish, opened in 1962, collapsed in 1988 because of metal fatigue; a replacement dish, 100 m/330 ft across, opened in 1999.

Grus
Conspicuous constellation of the Southern Hemisphere, represented as a crane.

Hadar, or Beta Centauri
Second brightest star in the constellation of Centaurus and the eleventh brightest in the sky. It is a blue-white giant star of magnitude 0.6, some 320 light-years from Earth. It is a binary star comprising two stars of magnitude 0.7 and 3.9.

Hale-Bopp, Comet
See Comet Hale-Bopp (C/1995 01).

Halley's comet
Comet that orbits the sun about every seventy-five years, named for English astronomer, mathematician, and physicist Edmond Halley, who calculated its orbit. It is the brightest and most conspicuous of the periodic comets. Recorded sightings go back more than two thousand years. It travels around the sun in the opposite direction to the planets. Its orbit is inclined at almost 20° to the main plane of the solar system and ranges between the orbits of Venus and Neptune. It will next reappear in 2061.

The comet was studied by space probes at its last appearance in 1986. The European probe *Giotto* showed that the nucleus of Halley's comet is a tiny and irregularly shaped chunk of ice, measuring 15 km/10 m long by 8 km/5 m wide, coated by a layer of very dark material, thought to be composed of carbon-rich compounds. This surface coating has a very low albedo (i.e., the fraction of incoming light it reflects), reflecting just 4 percent of the light it receives from the sun. Although the comet is one of the darkest objects known, it has a glowing head and tail produced by jets of gas from fissures in the outer dust layer. These vents cover 10 percent of the total surface area and become active only when exposed to the sun. The force of these jets affects the speed of the comet's travel in its orbit.

Hayashi track
A path on the Hertzsprung-Russell diagram taken by protostars as they emerge from the clouds of dust and gas out of which they were born. A protostar appears on the right (cool) side of the Hertzsprung-Russell diagram and follows a Hayashi track until it arrives on the main sequence, where hydrogen burning can start. It is named for Japanese astrophysicist Chushiro Hayashi, who studied the theory of protostars in the 1960s.

heat death
In cosmology, a possible fate of the universe in which it continues expanding indefinitely while all of the stars burn out and no new ones are formed.

helioseismology
Study of the sun's structure by analyzing vibrations and monitoring effects on the sun's surface.

heliosphere
Region of space through which the solar wind flows outward from the sun. The heliopause is the boundary of this region, believed to lie about 100 astronomical units from the sun, where the flow of the solar wind merges with the interstellar gas.

Hercules
In astronomy, the fifth-largest constellation, lying in the Northern Hemisphere. Despite its size, it contains no prominent stars. Its most important feature is the best example in the Northern Hemisphere of a globular cluster of stars, 22,500 light-years from Earth, which lies between Eta and Zeta Herculis.

Hertzsprung-Russell diagram
A graph on which the surface temperatures of stars are plotted against their luminosities. Most stars, including the sun, fall into a narrow band called the main sequence. When a star grows old, it moves from the main sequence to the upper right part of the graph, into the area of the giants and the supergiants. At the end of its life, as the star shrinks to become a white dwarf, it moves again, to the bottom left area. The diagram is named for Danish astronomer Ejnar Hertzsprung (1873–1967) and U.S. astronomer Henry Russell, who independently devised it (1911–1913).

HII region
A region of extremely hot ionized hydrogen, surrounding one or more hot stars, visible as a bright patch or emission nebula in the sky. The gas is ionized by the intense ultraviolet radiation from the stars within it. HII regions are often associated with interstellar clouds in which new stars are being born. An example is the Orion Nebula. The region takes its name from a spectroscopic notation in which HI represents neutral hydrogen (H) and HII represents ionized hydrogen (H+).

Hipparcos (acronym for *High Precision Parallax Collecting Satellite*)
Satellite launched by the European Space Agency in 1989. Named after the Greek astronomer Hipparchus, it is the world's first astrometry satellite and is providing precise positions, distances, colors, brightnesses, and apparent motions for more than 100,000 stars.

horizon
In astronomy, the great circle dividing the visible part of the sky from the part hidden by Earth.

Horologium
Inconspicuous constellation of the Southern Hemisphere, represented as a clock.

HOTOL (horizontal takeoff and landing)

Reusable hypersonic space plane invented in 1983 by British engineer Alan Bond but never put into production.

Hubble classification

A scheme for classifying galaxies according to their shapes, originally devised by U.S. astronomer Edwin Hubble in the 1920s.

Elliptical galaxies are classed from type E0 to E7, the figures denoting the degree of ellipticity. An E0 galaxy appears circular to an observer, while an E7 is highly elliptical (this is based on the apparent shape; the true shape, distorted by foreshortening, may be quite different). Spiral galaxies are classed as type Sa, Sb, or Sc: Sa is a tightly wound spiral with a large central bulge, Sc is loosely wound with a small bulge, and Sb is in between. Intermediate types are denoted by Sab or Sbc. Barred spiral galaxies, which have a prominent bar across their centers, are similarly classed as type SBa, SBb, or SBc with intermediates SBab or SBbc. Lenticular galaxies, which have no spiral arms, are classed as type S0. Irregular galaxies, type Irr, can be subdivided into Irr I, which resemble poorly formed spirals, and Irr II, which are otherwise

The Hubble classification was once believed to reveal an evolutionary sequence (from ellipticals to spirals), but this is now known not to be the case. Our own Milky Way galaxy is a spiral classified as type Sb or Sc but may have a bar.

Hubble Space Telescope (HST)

Space-based astronomical observing facility, orbiting Earth at an altitude of 610 km/380 mi. It consists of a 2.4-m/94-in telescope and four complementary scientific instruments, is roughly cylindrical, 13 m/43 ft long, and 4 m/13 ft in diameter, with two large solar panels. HST produces a wealth of scientific data and allows astronomers to observe the birth of stars, find planets around neighboring stars, follow the expanding remnants of exploding stars, and search for black holes in the center of galaxies. HST is a cooperative program between the European Space Agency (ESA) and the U.S. National Aeronautics and Space Administration (NASA) and is the first spacecraft specifically designed to be serviced in orbit as a permanent space-based observatory. It was launched in 1990.

By having a large telescope above Earth's atmosphere, astronomers are able to look at the universe with unprecedented clarity. Celestial observations by HST are unhampered by clouds and other atmospheric phenomena that distort and attenuate starlight. In particular, the apparent twinkling of starlight caused by density fluctuations in the atmosphere limits the clarity of ground-based telescopes. HST performs at least ten times better than such telescopes and can see almost back to the edge of the universe and to the beginning of time.

Before HST could reach its full potential, a flaw in the shape of its main mirror, discovered two months after the launch, had to be corrected. In 1993, as part of a planned servicing and instrument-upgrade mission, U.S. National Aeronautics and Space Administration (NASA) astronauts aboard the space shuttle *Endeavor* installed a set of corrective lenses to compensate for the error in the mirror figure. COSTAR (corrective optics space telescope axial replacement), a device containing ten coin-size mirrors, now feeds a corrected image from the main mirror to three of the HST's four scientific instruments. HST is also being used to detail the distribution of dust and stars in nearby galaxies, watch the collisions of galaxies in detail, infer the evolution of galaxies, and measure the age of the universe.

In December 1995, HST was trained on an "empty" area of sky near the Big Dipper, now termed the Hubble Deep Field. Around fifteen hundred galaxies, mostly new discoveries, were photographed.

Two new instruments were added in February 1997. The Near Infrared Camera and Multi-Object Spectrometer (NICMOS) enables Hubble to see things even farther away (and, therefore, older) than ever before. The Space Telescope Imaging Spectograph works thirty times faster than its predecessor, as it can gather information about different stars at the same time. Three new cameras had to be fitted shortly afterward as one of the original ones were found to be faulty.

In May 1997, three months after astronauts installed the new equipment, U.S. scientists reported that Hubble had made an extraordinary finding. Within twenty minutes of searching, it discovered evidence of a black hole 300 million times the mass of the sun. It is located in the middle of galaxy M84, about 50 million light-years from Earth. Further findings in December 1997 concerned different shapes of dying stars. Previously, astronomers had thought that most stars die with a round shell of burning gas expanding into space. The photographs taken by HST show shapes such as pinwheels and jet exhaust. This may be indicative of how the sun will die.

The HST cost $2.5 billion—five times the original estimate—and was launched seven years late by the space shuttle *Discovery* in April 1990. Its instruments include a wide field/planetary camera designed to gather the sharpest astronomical images ever. The faint-object camera was built by ESA and uses an image intensifier to image the faintest object resolvable by the telescope. The faint-object spectrometer (FOS) measures spectra in a wide range of light, from ultraviolet to near infrared. The Goddard high-resolution spectrograph is similar to the FOS but is dedicated to ultraviolet and infrared astronomy.

Hubble's constant

A measure of the rate at which the universe is expanding, named for U.S. astronomer Edwin Hubble. Observations previously suggested that galaxies are moving apart at a rate of 50–100 kps/30–60 mps for every million parsecs of distance. This means that the universe, which began at one point according to the Big Bang theory, is between ten billion and twenty billion years old. Observations by the Hubble Space Telescope in 1996 produced a revised constant of 73 kps/45 mps.

Hubble's Law

The law that relates a galaxy's distance from us to its speed of recession as the universe expands, announced in 1929 by U.S. astronomer Edwin Hubble. He found that galaxies are moving apart at speeds that increase in direct proportion to their distance apart. The rate of expansion is known as Hubble's constant.

Hyades

V-shaped cluster of stars that forms the face of the bull in the constellation Taurus. It is 150 light-years away from Earth and contains more than two hundred stars, although only about a dozen are visible to the naked eye. The Hyades is a much older cluster than the Pleiades, which is also in the constellation Taurus, for not only have some of the brighter stars evolved into red giants, some have gone even further and are now white dwarfs. Aldebaran, or Alpha Tauri, which marks the eye of the bull and which appears to be in the middle of the cluster, is not actually a member of the cluster. It is only 65 light-years away.

The Hyades is sometimes described as a "moving cluster," as the proper motions of its members are large, are very nearly the same, and stand out conspicuously from those of the surrounding stars. They appear to converge and, thus, provide an independent geometrical method of estimating the cluster's distance from Earth. This is particularly important because the observed relationship between the absolute magnitudes and the colors of its constituent members can be used to estimate the distances to more remote clusters, which, in turn, provide the scale with which all large intergalactic and extragalactic distances are measured.

Hydra

The largest constellation, winding across more than a quarter of the sky between Cancer and Libra in the Southern Hemisphere. Hydra is named for the multiheaded sea serpent slain by Hercules. Despite its size, it is not prominent; its brightest star is second-magnitude Alphard. It is sometimes identified as the River Nile or Jordan but is more usually represented as a snake trailing along the ground bearing on its back a cup (Crater) and near its tail a seated crow or raven (Corvus).

hydrogen burning

In astronomy, any of several processes by which hydrogen is converted to helium by nuclear fusion in the core of a star. In the sun, the main process is the proton-proton chain, while, in heavier stars, the carbon cycle is more important. In both processes, four protons are converted to a helium nucleus with the emission of positrons, neutrinos, and gamma rays. The temperature must exceed several million K for hydrogen burning to start, and the least massive stars (brown dwarfs) never become hot enough.

Hydrus

Constellation of the south polar region, represented as a water snake. It is quite separate from the equatorial constellation Hydra. Beta Hydri is the nearest bright star to the south pole.

Icarus

In astronomy, an Apollo asteroid 1.5 km/1 mi in diameter, discovered in 1949 by German-born U.S. astronomer Walter Baade. It orbits the sun every 409 days at a distance of 28 million km/18 million mi (0.19–2.0 astronomical units). It was the first asteroid known to approach the sun closer than does the planet Mercury. In 1968, it passed 6 million km/3.7 million mi from Earth.

inclination
Angle between the ecliptic and the plane of the orbit of a planet, asteroid, or comet. In the case of satellites orbiting a planet, it is the angle between the plane of orbit of the satellite and the equator of the planet.

Indus
Inconspicuous constellation of the south polar region, represented as a Native American.

inferior planet
A planet (Mercury or Venus) whose orbit lies between that of Earth and the sun.

inflation
In cosmology, a phase of extremely fast expansion thought to have occurred within 10^{-32} seconds of the Big Bang and in which almost all of the matter and energy in the universe was created. The inflationary model based on this concept accounts for the density of the universe being very close to the critical density, the smoothness of the cosmic background radiation, and the homogeneous distribution of matter in the universe. Inflation was proposed by U.S. astrophysicist Alan Guth in the early 1980s.

infrared astronomy
Study of infrared radiation produced by relatively cool gas and dust in space, as in the areas around forming stars. In 1983, the U.S.-Dutch-British *Infrared Astronomy Satellite* (IRAS) surveyed almost the entire infrared sky. It found five new comets, thousands of galaxies undergoing bursts of star formation, and the possibility of planetary systems forming around several dozen stars.

Planets and gas clouds emit their light in the far- and mid-infrared region of the spectrum. The *Infrared Space Observatory* (ISO), launched in November 1995 by the European Space Agency, observed a broad wavelength (3–200 micrometers) in these regions. It was ten thousand times more sensitive than IRAS and searched for brown dwarfs (cool masses of gas smaller than the sun).

Infrared Astronomy Satellite (IRAS)
Joint U.S.-British-Dutch satellite launched in January 1983 to survey the sky at infrared wavelengths, studying areas of star formation, distant galaxies, and possible embryo planetary systems around other stars and discovering five new comets in our own solar system. It operated for ten months.

Infrared Space Observatory (ISO)
Orbiting telescope with a 60-cm-/24-in-diameter mirror. It was launched November 17, 1995, by the European Space Agency and spent until May 1998 in an

elongated orbit, at a range from Earth of 1,000–70,500 km/620–43,800 mi and keeping it as much as possible outside the radiation belts that swamped its detectors. The ISO made the first-ever discovery of water vapor from a source beyond the solar system (in planetary nebula NGC 2027); traced the spiral arms of the Whirlpool galaxy and detected sites of star formation there; and obtained the first comprehensive spectrum of Saturn's atmosphere.

infrared telescope

A telescope designed to receive electromagnetic waves in the infrared part of the spectrum. Infrared telescopes are always reflectors (glass lenses are opaque to infrared waves) and are normally of the Cassegrain-telescope type.

Since all objects at normal temperatures emit strongly in the infrared, careful design is required to ensure that the weak signals from the sky are not swamped by radiation from the telescope itself. Infrared telescopes are sited at high mountain observatories above the obscuring effects of water vapor in the atmosphere. Modern large telescopes are often designed to work equally well in both visible and infrared light.

Intelsat

Acronym for International Telecommunications Satellite Organization, established in 1964 to operate a worldwide system of communications satellites. In 1999, it had 143 member nations and nineteen satellites in orbit. Its headquarters are in Washington, D.C. Intelsat satellites are stationed in geostationary orbit (maintaining their positions relative to Earth) over the Atlantic, Pacific, and Indian Oceans. The first Intelsat satellite was *Early Bird,* launched in 1965.

interferometry

Any of several techniques used in astronomy to obtain high-resolution images of astronomical objects.

International Ultraviolet Explorer (IUE)

Joint U.S. National Aeronautics and Space Administration (NASA) and European Space Agency (ESA) orbiting ultraviolet telescope with a 45-cm/18-in mirror, launched in 1978 to provide data on ultraviolet sources in space. It was switched off in September 1996.

interplanetary matter

Gas and dust thinly spread through the solar system. The gas flows outward from the sun as the solar wind. Fine dust lies in the plane of the solar system, scattering sunlight to cause the zodiacal light. Swarms of dust shed by comets enter Earth's atmosphere to cause meteor showers.

interstellar cirrus

Wispy cloudlike structures discovered in the mid-1980s by the *Infrared Astronomy Satellite* (IRAS) and believed to be the remains of dust shells blown into space from cool giant or supergiant stars.

interstellar matter

Medium of electrons, ions, atoms, molecules, and dust grains that fills the space between stars in our own and other galaxies. More than one hundred different types of molecule exist in gas clouds in Earth's galaxy. Most have been detected by their radio emissions, but some have been found by the absorption lines they produce in the spectra of starlight. The most complex molecules, many of them based on carbon, are found in the dense clouds in which stars are forming. They may be significant for the origin of life elsewhere in space.

It is only since the mid-twentieth century that scientists have realized that there is sufficient interstellar matter to have significant effects and that its extent largely determines the form and development of a galaxy. It is most easily observable in the radio region of the spectrum but was first detected optically. Condensations of such matter are visible as nebulae, while, over large parts of the sky, it dims, reddens, and polarizes the light of distant stars. It also causes a number of characteristic absorption lines in their spectra.

Early radio observations showed the general extent of interstellar matter; further observations plotted the distribution of its most abundant constituent, neutral hydrogen atoms. Later radio observations located hydroxyl, helium, water, ammonia, and many other molecules, some of them quite complex.

Interstellar matter is not smoothly distributed but occurs in dense and cold clouds. Its fundamental properties are largely determined by the hydrogen component. By mass, helium is 20–30 percent as abundant as hydrogen. All of the other elements together do not amount to more than 3–5 percent.

Their presence, particularly that of oxygen, nitrogen, carbon, sulfur, and iron, is important, as they help maintain the thermal balance of interstellar gas in the presence of radiation. Interstellar gas can be divided into neutral hydrogen (HI) and ionized hydrogen (HII) regions, though the two are, in many cases, intermingled. HII regions are emission nebulae surrounding hot stars. The term was introduced by Swedish astronomer Bengt Strömgren, who showed that a hot star can completely ionize any gas surrounding it out to a certain distance and that the boundary between the ionized and the neutral gas is quite sharp.

Submicron dust grains, which form about 1 percent by mass of the interstellar matter, also play an important role. They help cool the gas and provide a surface on which molecules can form. As they are thoroughly mixed with the gas, they can be used to trace the structural details of the gas clouds. It is thought that interstellar grains are composed of graphite, silicate, and ice. Iron and silicate carbide have also been proposed. These grains can form in many ways—for example, in the photosphere of cool red giants and during supernovae explosions.

Io

The third-largest moon of the planet Jupiter, 3,630 km/2,260 mi in diameter, orbiting in 1.77 Earth days at a distance of 422,000 km/262,000 mi. It is the most volcanically active body in the solar system, covered by hundreds of vents that erupt not lava but sulfur, giving Io an orange-colored surface.

In July 1995, the Hubble Space Telescope revealed the appearance of a 320-km/200-mi yellow spot on the surface of Io, located on the volcano Ra Patera.

Though clearly volcanic in origin, astronomers are unclear as to the exact cause of the new spot. Using data gathered by the U.S. spacecraft *Galileo,* U.S. astronomers concluded in 1996 that Io has a large metallic core. *Galileo* also detected a 10-megawatt beam of electrons flowing between Jupiter and Io. In 1997, instruments aboard *Galileo* measured the temperature of Io's volcanoes and detected a minimum temperature of 1,800 K (1,500°C/2.732°F); in comparison, Earth's hottest volcanoes reach only about 1,600 K (1,300°C/1,372°F).

IRAS
See Infrared Astronomy Satellite (IRAS).

irregular galaxy
A class of galaxy with little structure, which does not conform to any of the standard shapes in the Hubble classification. The two satellite galaxies of the Milky Way, the Magellanic Clouds, are both irregulars. Some galaxies previously classified as irregulars are now known to be normal galaxies distorted by tidal effects or undergoing bursts of star formation.

IUE
See International Ultraviolet Explorer (IUE).

Jeans mass
The mass that a cloud (or part of a cloud) of interstellar gas must have before it can contract under its own weight to form a protostar. The Jeans mass is an expression of the Jeans criterion, which says that a cloud will contract when the gravitational force tending to draw material toward its center is greater than the opposing force due to gas pressure. It is named for British mathematician James Hopwood Jeans, whose work focused on the kinetic theory of gases and the origins of the cosmos.

jet
In astronomy, a narrow luminous feature seen protruding from a star or a galaxy and representing a rapid outflow of material.

Jet Propulsion Laboratory
U.S. National Aeronautics and Space Administration (NASA) installation at Pasadena, California, operated by the California Institute of Technology. It is the command center for NASA's deep-space probes, such as the *Voyager, Magellan,* and *Galileo* missions, with which it communicates via the Deep Space Network of radio telescopes at Goldstone, California; Madrid, Spain; and Canberra, Australia.

Jodrell Bank
Site in Cheshire, England, of the Nuffield Radio Astronomy Laboratories of the University of Manchester. Its largest instrument is the 76-m/250-ft radio dish (the Lovell Telescope), completed in 1957 and modified in 1970. A 38-x-25-m/125-x-

82-ft elliptical radio dish was introduced in 1964, capable of working at shorter wave lengths. These radio telescopes are used in conjunction with five smaller dishes up to 230 km/143 mi apart in an array called MERLIN (Multi-Element Radio-Linked Interferometer Network) to produce detailed maps of radio sources.

Johnson Space Center

U.S. National Aeronautics and Space Administration (NASA) installation at Houston, Texas, home of mission control for crewed space missions. It is the main center for the selection and training of astronauts.

Julian date

In astronomy, a measure of time in which days are numbered consecutively from noon Greenwich Mean Time (GMT) on January 1, 4713 B.C. It is useful when astronomers wish to compare observations made over long time intervals. The Julian date (J.D.) at noon on January 1, 2000, will be 2451545.0. The modified Julian date (MJD), defined as MJD = J.D. - 2400000.5, is more commonly used since the date starts at midnight GMT and the smaller numbers are more convenient.

Jupiter

The fifth planet from the sun and the largest in the solar system, with a mass equal to 70 percent of all of the other planets combined and 318 times that of Earth's. It is largely composed of hydrogen and helium, liquefied by pressure in its interior, and probably with a rocky core larger than Earth. Its main feature is the Great Red Spot, a cloud of rising gases, 14,000 km/8,500 mi wide and 30,000 km/20,000 mi long, revolving counterclockwise.

mean distance from the sun:
778 million km/484 million mi

equatorial diameter:
142,800 km/88,700 mi

rotation period:
0.41 Earth day

year:
(complete orbit) 11.86 Earth years

atmosphere:
Consists of clouds of white ammonia crystals, drawn out into belts by the planet's high speed of rotation (the fastest of any planet). Darker orange and brown clouds at lower levels may contain sulfur, as well as simple organic compounds. Farther down still, temperatures are warm, a result of heat left over from Jupiter's formation, and it is this heat that drives the turbulent weather

patterns of the planet. In 1995, the U.S. *Galileo* probe revealed that Jupiter's atmosphere consists of 0.2 percent water, less than previously estimated.

The Great Red Spot was first observed in 1664. Its top is higher than the surrounding clouds; its color is thought to be due to red phosphorus. Jupiter's strong magnetic field gives rise to a large surrounding magnetic "shell," or magnetosphere, from which bursts of radio waves are detected. The Southern Equatorial Belt in which the Great Red Spot occurs is subject to unexplained fluctuation. In 1989, it sustained a dramatic and sudden fading.

Comet Shoemaker-Levy 9 crashed into Jupiter in July 1994. Impact zones were visible but are not likely to remain.

surface:
Although largely composed of hydrogen and helium, Jupiter probably has a rocky core larger than Earth.

satellites:
Jupiter has sixteen known moons. The four largest moons, Io, Europa (which is about the size of our moon), Ganymede, and Callisto, are the Galilean satellites, discovered in 1610 by Italian astronomer and physicist Galileo (Ganymede, which is larger than Mercury, is the largest moon in the solar system). Three small moons were discovered in 1979 by the U.S. *Voyager* space probes, as was a faint ring of dust around Jupiter's equator 55,000 km/34,000 mi above the cloud tops.

Kagoshima Space Center

Headquarters of Japan's Institute of Space and Astronautical Science (ISAS), situated on South Kyushu Island. ISAS is responsible for the development of satellites for scientific research; other aspects of the space program fall under the National Space Development Agency, which runs the Tanegashima Space Center. Japan's first satellite was launched from Kagoshima in 1970. By 1988, ISAS had launched seventeen satellites and space probes.

Keck I Telescope

World's largest optical telescope, situated on Mauna Kea, Hawaii. It has a primary mirror 10 m/33 ft in diameter, unique in that it consists of thirty-six hexagonal sections, each controlled and adjusted by a computer to generate single images of the objects observed; it weighs 300 metric tons/331 tons. It received its first images in 1990 and became fully functional in 1992. An identical telescope next to it, named Keck II, became operational in 1996. Both telescopes are jointly owned by the California Institute of Technology, Pasadena, and the University of California/Lick Observatory.

Kennedy Space Center

U.S. National Aeronautics and Space Administration (NASA) launch site on Merritt Island, near Cape Canaveral, Florida, used for *Apollo* and space-shuttle launches. The first flight to land on the moon (1969) and *Skylab*, the first orbiting laboratory (1973), were launched here. The center is dominated by the Vehicle Assembly Building, 160 m/525 ft tall, used for assembly of *Saturn* rockets and space shuttles.

Kepler's laws

Three laws of planetary motion formulated in 1609 and 1619 by German mathematician and astronomer Johannes Kepler: (1) the orbit of each planet is an ellipse, with the sun at one of the foci; (2) the radius vector of each planet sweeps out equal areas in equal times; and (3) the squares of the periods of the planets are proportional to the cubes of their mean distances from the sun. Kepler derived the laws after exhaustive analysis of numerous observations of the planets, especially Mars, made by Danish astronomer Tycho Brahe without telescopic aid. British physicist and mathematician Isaac Newton later showed that Kepler's laws were a consequence of the theory of universal gravitation.

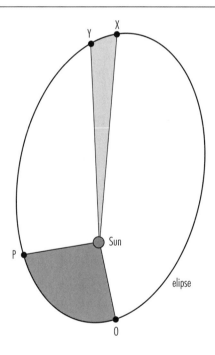

Kirkwood gaps

Regions of the asteroid belt, between Mars and Jupiter, where there are relatively few asteroids. The orbital periods of particles in the gaps correspond to simple fractions, especially 1/3, 2/5, 3/7, and 1/2, of the orbital period of Jupiter, indicating that they are caused by the gravitational influence of the larger planet. The gaps are named for Daniel Kirkwood, the nineteenth-century U.S. astronomer who first drew attention to them.

Kitt Peak National Observatory
Observatory in the Quinlan Mountains near Tucson, Arizona, operated by AURA (Association of Universities for Research into Astronomy) in agreement with the National Science Foundation of the United States. It is a division of the National Optical Astronomy Observatories. Its main telescopes are the 4-m/158-in Mayall reflector, opened in 1973, and the McMath-Pierce Solar Telescope, opened in 1962, the world's largest of its type. Among numerous other telescopes on the site are a 2.3-m/90-in reflector owned by the Steward Observatory of the University of Arizona and a 3.5-m/138-in reflecting telescope, opened in 1994, owned by the WIYN consortium, which comprises the University of Wisconsin, Indiana University, Yale University, and the National Optical Astronomy Observatories (NOAO).

Kourou
Second-largest town of French Guiana, site of the Guiana Space Center of the European Space Agency (ESA). Situated near the equator, it is an ideal site for launches of satellites into geostationary orbit.

Kuiper belt
Ring of small, icy bodies orbiting the sun beyond the outermost planet. The Kuiper belt, named for Dutch-born U.S. astronomer Gerard Kuiper, who proposed its existence in 1951, is thought to be the source of comets that orbit the sun with periods of less than two hundred years. The first member of the Kuiper belt was seen in 1992. In 1995, the first comet-size objects were discovered; previously, the only objects found had diameters of at least 100 km/62 mi (comets generally have diameters of less than 10 km/6.2 mi).

Two new objects were discovered in the Kuiper belt in 1996. The first, 1996 TL66, is 500 km/300 mi in diameter and has an irregular orbit that takes it four to six times farther from the sun than Neptune. The second, 1996 RQ20, is slightly smaller, with an orbit that takes it about three times farther from the sun than Neptune. The orbits of both are at an angle of 20° to the plane of the solar system.

Lacerta
Inconspicuous constellation of the Northern Hemisphere, represented as a lizard.

Lagrangian points
Five locations in space where the centrifugal and gravitational forces of two bodies neutralize each other; a third, less massive body located at any one of these points will be held in equilibrium with respect to the other two. Three of the points, L1–L3, lie on a line joining the two large bodies. The other two points, L4 and L5, which are the most stable, lie on either side of this line. Their existence was predicted in 1772 by French mathematician Joseph Louis Lagrange. The Trojan asteroids lie at Lagrangian points L4 and L5 in Jupiter's orbit around the sun.

Clouds of dust and debris may lie at the Lagrangian points of the moon's orbit around Earth.

Landsat
Series of U.S. satellites used for monitoring Earth's resources. The first was launched in 1972.

Las Campanas Observatory
Site in Chile of the 2.5-m/100-in Irénée du Pont telescope and also of a 1-m/39-in reflector, operated by the Carnegie Institution of Washington, D.C., opened in 1977.

lens, gravitational
See gravitational lensing.

lenticular galaxy
A lens-shaped galaxy with a large central bulge and flat disk but no discernible spiral arms.

Leo
Zodiacal constellation in the Northern Hemisphere, represented as a lion. The sun passes through Leo from mid-August to mid-September. Its brightest star is first-magnitude Regulus at the base of a pattern of stars called the Sickle. In astrology, the dates for Leo are about July 23–August 22.

Leo Minor
Inconspicuous constellation of the Northern Hemisphere, represented as a little lion.

Lepus
Small constellation on the celestial equator (celestial sphere), represented as a hare. It is said to be running from Orion and his dogs, Canis Major and Canis Minor.

Libra
Faint zodiacal constellation on the celestial equator adjoining Scorpius and represented as the scales of justice. The sun passes through Libra during November. The constellation was once considered to be a part of Scorpius, seen as the scorpion's claws. In astrology, the dates for Libra are about September 23–October 23.

libration
A slight, apparent wobble in the rotation of the moon due to its variable speed of rotation and the tilt of its axis.

Generally, the moon rotates on its axis in the same time as it takes to complete one orbit, causing it to keep one face turned permanently toward Earth. Its speed in orbit varies, however, because its orbit is not circular but elliptical, so at times the moon's axial rotation appears to get either slightly ahead of or slightly behind its orbital motion, making part of the "dark side" of the moon visible around the east and west edges. This is known as libration in longitude. Libration in latitude occurs because the moon's axis is slightly tilted with respect to its orbital plane, so we can see over the north and south poles. In combination, these effects mean that a total of 59 percent of the moon's surface is visible, rather than just 50 percent if libration did not occur.

Lick Observatory
Observatory of the University of California at Mount Hamilton, California. Its main instruments are the 3.04-m/120-in Shane reflector, opened in 1959, and a 91-cm/36-in refractor, opened in 1988, the second-largest refractor in the world.

light curve
A graph showing how the brightness of an astronomical object varies with time. Analysis of the light curves of variable stars, for example, gives information about the physical processes causing the variation.

light second
Unit of length, equal to the distance traveled by light in one second. It is equal to 2.997925×108 m/9.835592×108 ft.

light-year
The distance traveled by a beam of light in a vacuum in one year, approximately 9.4605×10^{12} km/5.9128×10^{12} trillion mi.

Little Dipper
Popular name for the most distinctive part of the constellation Ursa Minor, the Little Bear.

Local Group
A cluster of about thirty galaxies that includes our own, the Milky Way. Like other groups of galaxies, the Local Group is held together by the gravitational attraction among its members and does not expand with the expanding universe. Its two largest galaxies are the Milky Way and the Andromeda galaxy; most of the others are small and faint.

lodestar, or loadstar
A star used in navigation or astronomy, often Polaris, the Pole Star.

Lowell Observatory
Astronomical observatory founded by U.S. astronomer Percival Lowell at Flagstaff, Arizona, with a 61-cm/24-in refractor, opened in 1896. The observatory now

operates other telescopes at a nearby site on Anderson Mesa, including the 1.83-m/72-in Perkins reflector of Ohio State and Ohio Wesleyan Universities.

luminosity, or brightness
In astronomy, the amount of light emitted by a star, measured in magnitudes. The apparent brightness of an object decreases in proportion to the square of its distance from the observer. The luminosity of a star or other body can be expressed in relation to that of the sun.

Lunar Alps
See Alps, Lunar.

Lunar Apennines
See Apennines, Lunar.

Lupus
Constellation of the Southern Hemisphere, represented as a wolf. Most of its brighter stars form part of Gould's belt.

Lynx
Inconspicuous constellation of the Northern Hemisphere.

Lyra
Small but prominent constellation of the Northern Hemisphere, represented as the harp of Orpheus. Its brightest star is Vega. Epsilon Lyrae is a system of four gravitationally linked stars. Beta Lyrae is an eclipsing binary. The Ring nebula, M57, is a planetary nebula.

MACHO
Abbreviation for massive astrophysical compact halo object, a component of the galaxy's dark matter. Most MACHOs are believed to be brown dwarfs, tiny failed stars with a mass of about 0.08 that of the sun, but they may also include neutron stars left behind after supernova explosions. MACHOs are identifiable when they move in front of stars, causing microlensing (magnification) of the star's light. Astronomers first identified MACHOs in 1993 and estimate that they account for 20 percent of the dark matter.

Magellan
U.S. National Aeronautics and Space Administration (NASA) space probe to Venus, launched in May 1989; it went into orbit around Venus in August 1990 to make a detailed map of the planet by radar. It revealed volcanoes, meteorite craters, and fold mountains on the planet's surface. *Magellan* mapped 99 percent of Venus. In October 1994, *Magellan* was instructed to self-destruct by entering the atmosphere around Venus, where it burned up.

Magellanic Clouds

The two galaxies nearest to our own. They are irregularly shaped and appear as detached parts of the Milky Way, in the southern constellations Dorado, Tucana, and Mensa.

The Large Magellanic Cloud spreads over the constellations of Dorado and Mensa. The Small Magellanic Cloud is in Tucana. The Large Magellanic Cloud is 169,000 light-years from Earth and about a third the diameter of our galaxy; the Small Magellanic Cloud, 180,000 light-years away, is about a fifth the diameter of our galaxy. They are named for the early-sixteenth-century Portuguese navigator Ferdinand Magellan, who first described them.

Being the nearest galaxies to ours (the distance to the Great Nebula in Andromeda is 2.2 million light-years), the Magellanic Clouds are especially useful for studying stellar populations and objects such as supergiant stars. It was in the Small Magellanic Cloud that the period-luminosity relationship for Cepheid variables was first established by U.S. astronomer Henrietta Leavitt in 1912.

magnetosphere

Volume of space surrounding a planet that is controlled by the planet's magnetic field and acts as a magnetic "shell." The Earth's magnetosphere extends 64,000 km/40,000 mi toward the sun but many times this distance on the side away from the sun.

The extension away from the sun is called the magnetotail. The outer edge of the magnetosphere is the magnetopause. Beyond this is a turbulent region, the magnetosheath, where the solar wind is deflected around the magnetosphere. Inside the magnetosphere, atomic particles follow Earth's lines of magnetic force. The magnetosphere contains the Van Allen radiation belts. Other planets have magnetospheres, notably Jupiter.

magnitude

In astronomy, measure of the brightness of a star or other celestial object. The larger the number denoting the magnitude, the fainter the object. Zero or first magnitude indicates some of the brightest stars. Still brighter are those of negative magnitude, such as Sirius, whose magnitude is –1.46. Apparent magnitude is the brightness of an object as seen from Earth; absolute magnitude is the brightness at a standard distance of 10 parsecs (32.616 light-years).

Each magnitude step is equal to a brightness difference of 2.512 times. Thus, a star of magnitude 1 is $(2.512)^5$, or one hundred times brighter than a sixth-magnitude star just visible to the naked eye. The apparent magnitude of the sun is –26.8, its absolute magnitude is +4.8.

main sequence

The part of the Hertzsprung-Russell diagram that contains most of the stars, including the sun. It runs diagonally from the top left of the diagram to the lower right. The most massive (hence, brightest) stars are at the top left, with the least massive (coolest) stars at the bottom right.

mare (plural maria)

Dark lowland plain on the moon. The name comes from the Latin for "sea," because these areas were once wrongly thought to be water.

Mariner spacecraft

Series of U.S. space probes that explored the planets Mercury, Venus, and Mars (1962–1975).

   Mariner 1 (to Venus, 1962) had a failed launch. *Mariner 2* (1962) made the first flyby of Venus, at 34,000 km/21,000 mi, confirmed the existence of solar wind, and measured Venusian temperature. *Mariner 3* (launched November 4, 1964) did not achieve its intended trajectory to Mars. *Mariner 4* (launched November 28, 1964) passed Mars at a distance of 9,800 km/6,100 mi (1965) and took photographs, revealing a dry, cratered surface. *Mariner 5* (1967) passed Venus at 4,000 km/2,500 mi and measured Venusian temperature, atmosphere, mass, and diameter. *Mariner 6 and 7* (1969) photographed Mars's equator and southern hemisphere, respectively, and also measured temperature, atmospheric pressure and composition, and diameter. *Mariner 8* (to Mars, 1971) had a failed launch. *Mariner 9* (1971) mapped the entire Martian surface and photographed Mars's moons. Its photographs revealed the changing of the polar caps and the extent of volcanism, canyons, and features, which suggested that there might once have been water on Mars. *Mariner 10* (1973–1974) took close-up photographs of Mercury and Venus and measured temperature, radiation, and magnetic fields.

   Mariner 11 and 12 were renamed *Voyager 1* and *2* (launched in 1977).

Mars

Fourth planet from the sun. Mars may approach Earth to within 54.7 million km/ 34 million mi. It is much smaller than Venus or Earth, with a mass 0.11 that of Earth. Mars is slightly pear shaped, with a low, level northern hemisphere, which is comparatively uncratered and geologically "young," and a heavily cratered "ancient" southern hemisphere.

   mean distance from the sun:
   227.9 million km/141.6 million mi

   equatorial diameter:
   6,780 km/4,210 mi

   rotation period:
   1.03 Earth days

   year:
   687 Earth days

   atmosphere:
   Carbon dioxide (95 percent), nitrogen (3 percent), argon (1.5 percent), and oxygen (0.15 percent). Red atmospheric dust from the surface whipped up by

winds of up to 450 kph/280 mph accounts for the light pink sky. The surface pressure is less than 1 percent of Earth's atmospheric pressure at sea level.

surface:

The landscape is a dusty, red, eroded lava plain. Mars has white polar caps (water ice and frozen carbon dioxide) that advance and retreat with the seasons. There are four enormous volcanoes near the equator, of which the largest is Olympus Mons, 24 km/15 mi high with a base 600 km/375 mi across, and a crater 65 km/40 mi wide. To the east of the four volcanoes lies a high plateau cut by a system of valleys, Valles Marineris, some 4,000 km/2,500 mi long, up to 200 km/120 mi wide, and 6 km/4 mi deep; these features are apparently caused by faulting and wind erosion. Recorded temperatures vary from –100°C/–148°F to 0°C/32°F.

satellites:

Two small satellites: Phobos and Deimos

The first human-made object to orbit another planet was the U.S. craft *Mariner 9,* which began orbiting Mars in November 1971. *Viking 1* and *Viking 2,* which landed on Mars, also provided much information. Studies in 1985 showed that enough water might exist to sustain prolonged missions by space crews.

In December 1996, U.S. National Aeronautics and Space Administration (NASA) launched the *Mars Pathfinder,* which made a successful landing on Mars in July 1997 on a flood plain called Ares Vallis. After initial technical problems, its rover, *Sojourner,* began to explore the Martian landscape and to transmit data back to Earth. Based on photographs from the *Mars Pathfinder,* NASA announced in July 1997 that the planet is rusting, that a supercorrosive force was eroding rocks on the surface due to iron oxide in the soil.

In May 1997, U.S. scientists announced that Mars was becoming colder and cloudier. Images from the Hubble Space Telescope showed that dust storms had covered areas of the planet that had been dark features in the early twentieth century, including one section as large as California.

The *Mars Global Surveyor,* launched November 7, 1996, entered Martian orbit in September 1997; its data revealed that Mars's magnetic field is only 0.00125 that of Earth.

Mars Global Surveyor

U.S. spacecraft that went into orbit around Mars September 12, 1997, to conduct a detailed photographic survey of the planet, commencing in March 1998. The spacecraft used a previously untried technique called aerobraking to turn its initially highly elongated orbit into a 400-km/249-mi circular orbit by dipping into the outer atmosphere of the planet.

Mars Observer

U.S. National Aeronautics and Space Administration (NASA) space probe launched in 1992 to orbit Mars and survey the planet, its atmosphere, and the polar caps over two years. The probe was also scheduled to communicate information from the robot vehicles delivered by Russia's *Mars 94* mission. The $980 million project miscarried, however, when the probe unaccountably stopped transmitting in August 1993, three days before it was due to drop into orbit.

Mars Pathfinder

U.S. spacecraft that landed in the Ares Vallis region of Mars on July 4, 1997. It carried a small six-wheeled roving vehicle called *Sojourner* that examined rock and soil samples around the landing site. *Mars Pathfinder* was the first to use air bags instead of retro-rockets to cushion the landing.

Marshall Space Flight Center

U.S. National Aeronautics and Space Administration (NASA) installation at Huntsville, Alabama, at which the series of *Saturn* rockets and the space-shuttle engines were developed. It also manages various payloads for the space shuttle, including the *Spacelab* space station.

massive astrophysical compact halo object

See MACHO.

Mathilde

Rocky carbon-rich asteroid that is 53 km/33 mi in length. It is dominated by a huge 25-km/15.5-mi crater.

Mauna Kea

Astronomical observatory in Hawaii, built on a dormant volcano at 4,200 m/ 13,784 ft above sea level. Because of its elevation high above clouds, atmospheric moisture, and artificial lighting, Mauna Kea is ideal for infrared astronomy. The first telescope on the site was installed in 1970.

Telescopes include the 2.24-m/88-in University of Hawaii reflector, which opened in 1970. In 1979, three telescopes were erected: the 3.8-m/150-in United Kingdom Infrared Telescope (UKIRT), which is also used for optical observations; the 3-m/120-in U.S. National Aeronautics and Space Administration (NASA) Infrared Telescope Facility (IRTF); and the 3.6-m/142-in Canada-France-Hawaii Telescope (CFHT), designed for optical and infrared work. The 15-m/50-ft Britain-Netherlands James Clerk Maxwell Telescope (JCMT) is the world's largest telescope specifically designed to observe millimeter wave radiation from nebulae, stars, and galaxies. The JCMT is operated via satellite links by astronomers in Europe. The world's largest optical telescopes, the Kecks I and II Telescopes, are also situated on Mauna Kea.

In 1996, the capacity of the JCMT was enhanced by the addition of SCUBA (Submillimeter Common-User Bolometer Array). SCUBA is a camera that comprises numerous detectors cooled to within one-tenth of a degree of absolute zero (0 K) and is the world's most sensitive instrument at the 0.3–1.0-mm/0.01–0.04-in wavelength.

Maunder minimum

The period (1645–1715) when sunspots were rarely seen and no aurorae (northern lights) were recorded. The Maunder minimum coincided with a time of unusually low temperature on Earth, known as the Little Ice Age, and is often taken as evidence that changes in solar activity can affect Earth's climate. The Maunder minimum is named for English astronomer E. W. Maunder, who drew attention to it.

McDonald Observatory

Observatory of the University of Texas on Mount Locke in the Davis Mountains, Texas. It is the site of a 2.72-m/107-in reflector, opened in 1969, and a 2.08-m/82-in reflector, opened in 1939.

Mensa

Faint constellation of the south polar region. Eighteenth-century French astronomer Nicolas Lacaille gave it the name Mons Mensae in honor of Table Mountain, "which had witnessed his nightly vigils and daily toils" at the observatory at the Cape of Good Hope.

Mercury

The closest planet to the sun. Its mass is 0.056 that of Earth. On its sunward side, the surface temperature reaches more than 400°C/752°F, but, on the "night" side, it falls to −170°C/−274°F.

mean distance from the sun:
58 million km/36 million mi

equatorial diameter:
4,880 km/3,030 mi

rotation period:
59 Earth days

year:
88 Earth days

atmosphere:
Mercury's small mass and high daytime temperatures make it impossible for an atmosphere to be retained.

surface:
Composed of silicate rock, often in the form of lava flows. In 1974, the U.S. space probe *Mariner 10* showed that Mercury's surface is cratered by meteorite impacts. Its largest-known feature is the Caloris Basin, 1,400 km/870 mi wide. There are also cliffs hundreds of kilometers long and up to 4 km/2.5 mi high, thought to have been formed by the cooling of the planet billions of years

ago. Inside is an iron core 75 percent of the planet's diameter, which produces a magnetic field 1 percent the strength of Earth's.

satellites:
None

Mercury Project
U.S. project to put a human in space in the one-seat *Mercury* spacecraft (1961–1963). The first two *Mercury* flights, on *Redstone* rockets, were short flights to the edge of space and back. The orbital flights, beginning with the third in the series (made by John Glenn), were launched by *Atlas* rockets.

MERLIN array
Acronym for Multi-Element Radio-Linked Interferometer Network; a radio-telescope network centered at Jodrell Bank in Cheshire, England.

Messier catalog
A catalog of 103 galaxies, nebulae, and star clusters (the Messier objects) published in 1784 by French astronomer Charles Messier. Catalog entries are denoted by the prefix "M." Well-known examples include M31 (the Andromeda galaxy), M42 (the Orion Nebula), and M45 (the Pleiades star cluster). Messier compiled the catalog to identify fuzzy objects that could be mistaken for comets. The list was later extended to 109.

meteor
A flash of light in the sky, popularly known as a shooting or falling star, caused by a particle of dust, a meteoroid, entering the atmosphere at speeds up to 70 kps/45 mps and burning up by friction at a height of around 100 km/60 mi. On any clear night, several sporadic meteors can be seen each hour.

Several times each year Earth encounters swarms of dust shed by comets, which give rise to a meteor shower. This appears to radiate from one particular point in the sky, after which the shower is named; for example, the Perseid meteor shower in August appears in the constellation Perseus. A brilliant meteor is termed a fireball. Most meteoroids are smaller than grains of sand. The Earth sweeps up an estimated 16,000 metric tons/17,637 tons of meteoric material every year.

meteorite
A piece of rock or metal from space that reaches the surface of Earth, the moon, or other body. Most meteorites are thought to be fragments from asteroids, although some may be pieces from the heads of comets. Most are stony, although some are made of iron, and a few have a mixed rock-iron composition.

Stony meteorites can be divided into two kinds: chondrites and achondrites. Chondrites contain chondrules, small spheres of the silicate minerals olivine and orthopyroxene, and make up 85 percent of meteorites. Achondrites do not contain chondrules. Meteorites provide evidence for the nature of the solar system and may be similar to Earth's core and mantle, neither of which can be observed directly.

Thousands of meteorites hit Earth each year, but most fall in the sea or in remote areas and are never recovered. The largest-known meteorite is one composed of iron, weighing 60 metric tons/66 tons, which lies where it fell in prehistoric times at Grootfontein, Namibia. Meteorites are slowed by Earth's atmosphere, but, if they are moving fast enough, they can form a crater on impact. Meteor Crater in Arizona, about 1.2 km/0.7 mi in diameter and 200 m/660 ft deep, is the site of a meteorite impact about fifty thousand years ago.

In August 1996, U.S. National Aeronautics and Space Administration (NASA) scientists revealed that a meteorite from Mars that was found in Antarctica in 1984 contained possible evidence of life. The fifteen-million-year-old rock, which entered Earth's atmosphere in the form of a meteorite twelve thousand years ago, was probably broken from the surface of Mars when the planet collided with a large object in space. The meteorite, ALH84001, weighs 1.9 kg/4.2 lb and was found to contain carbonate globules, on the surface of which are minute round or long and thin microorganisms. These microfossils are one hundred times smaller than the smallest-known Earth microfossils. The globules also contain magnetite and iron-sulfide particles comparable in shape and quantity to those produced by Earth microorganisms. Critics claim the meteorite was contaminated. In 1998, there were twelve meteorites believed to have come from Mars.

meteoroid

Chunk of rock in interplanetary space. There is no official distinction between meteoroids and asteroids, except that the term "asteroid" is generally reserved for objects larger than 1.6 km/1 mi in diameter, whereas meteoroids can range anywhere from pebble size up. Meteoroids are believed to result from the fragmentation of asteroids after collisions. Some meteoroids enter Earth's atmosphere, and their fiery trails are called meteors. If they fall to Earth, they are named meteorites.

Microscopium

Inconspicuous constellation of the Southern Hemisphere, represented as a microscope.

Milky Way

Faint band of light crossing the night sky, consisting of stars in the plane of our galaxy. The name Milky Way is often used for the galaxy itself. It is a spiral galaxy, 100,000 light-years in diameter and 2,000 light-years thick, containing at least 100 billion stars. The sun is in one of its spiral arms, about 25,000 light-years from the center, not far from its central plane.

The densest parts of the Milky Way, toward the galaxy's center, lie in the constellation Sagittarius. In places, the Milky Way is interrupted by lanes of dark dust that obscure light from the stars beyond, such as the Coalsack Nebula in Crux (the Southern Cross). It is because of these that the Milky Way is irregular in width and appears to be divided in two between Centaurus and Cygnus.

The Milky Way passes through the constellations of Cassiopeia, Perseus, Auriga, Orion, Canis Major, Puppis, Vela, Carina, Crux, Centaurus, Norma, Scorpius, Sagittarius, Scutum, Aquila, and Cygnus.

Mills Cross
Type of radio telescope consisting of two rows of aerials at right angles to each other, invented in 1953 by Australian radio astronomer Bernard Mills. The cross shape produces a narrow beam useful for pinpointing the positions of radio sources.

Mimosa, or Beta Crucis
Second brightest star in the constellation of Crux, marking one of the four corners of the Southern Cross and the nineteenth brightest star in the sky. It is a blue-white giant star some 420 light-years from Earth.

minor planet
See asteroid, or minor planet.

Mir
Russian for "peace" or "world"; name of the Soviet (now Russian) space station, the core of which was launched on February 20, 1986. It is intended to be a permanently occupied space station.

Mir weighs almost 21 metric tons/23 tons, is approximately 13.5 m/44 ft long, and has a maximum diameter of 4.15 m/13.6 ft. It carries a number of improvements over the earlier *Salyut* series of space stations, including six docking ports; four of these can have scientific and technical modules attached to them. The first of these was the *Kvant* (quantum) astrophysics module, launched in 1987. This had two main sections: a main experimental module and a service module that would be separated in orbit. The experimental module was 5.8 m/19 ft long and had a maximum diameter matching that of *Mir*. When attached to the *Mir* core, *Kvant* added a further 40 m³/1,413 ft³ of working space to that already there. Among the equipment carried by *Kvant* were several X-ray telescopes and an ultraviolet telescope.

In June 1995, the U.S. space shuttle *Atlantis* docked with *Mir,* exchanging crew members. Since then, several U.S. astronauts have spent time on *Mir.* On December 6, 1996, a small wheat crop was harvested aboard the Soviet space station, the first successful cultivation of a plant from seed in space. In 1997, *Mir* suffered a series of problems, culminating in its collision with an unmanned cargo ship in June.

Mira, or Omicron Ceti
Brightest long-period pulsating variable star, located in the constellation Cetus. Mira was the first star discovered to vary periodically in brightness. It has a periodic variation between third or fourth magnitude and ninth magnitude over an average period of 331 days. It can sometimes reach second magnitude and once almost attained first magnitude, in 1779. At times, it is easily visible to the naked

eye, being the brightest star in that part of the sky, while at others it cannot be seen without a telescope.

Mizar, or Zeta Ursae Majoris

Second-magnitude star in Ursa Major, where it marks the middle of the handle of the plow. It was discovered to be a double star in 1650 and was the first star found to be double by using a telescope. Its brighter component was the first spectroscopic binary to be discovered, by U.S. astronomer Edward Pickering, in 1889. The second component is also a spectroscopic binary. The Mizar system, together with Alcor, a fourth-magnitude star, form a visual double star that the North American Indians called "the Horse and Rider."

molecular cloud

An enormous cloud of cool interstellar dust and gas containing hydrogen molecules and more complex molecular species. Giant molecular clouds (GMCs), about a million times as massive as the sun and up to 300 light-years in diameter, are regions in which stars are being born. The Orion Nebula is part of a GMC.

Monoceros

Constellation on the celestial equator, represented as a unicorn, which, though it includes no bright stars, lies in the Milky Way and contains many faint stars, clusters, and nebulae.

moon

Any natural satellite that orbits a planet. Mercury and Venus are the only planets in the solar system that do not have moons.

moon (Earth's)

Natural satellite of Earth, 3,476 km/2,160 mi in diameter, with a mass 0.012 (approximately one-eightieth) that of Earth. Its surface gravity is only 0.16 (one-sixth) that of Earth. Its average distance from Earth is 384,400 km/238,855 mi, and it orbits in a west-to-east direction every 27.32 days (the sidereal month). It spins on its axis with one side permanently turned toward Earth. The moon has no atmosphere and was thought to have no water until ice was discovered on its surface in 1998.

phases
The moon is illuminated by sunlight and goes through a cycle of phases of shadow, waxing from new (dark) via first quarter (half moon) to full and waning back again to new every 29.53 days (the synodic month, also known as a lunation). On its sunlit side, temperatures reach 110°C/230°F, but, during the two-week lunar night, the surface temperature drops to –170°C/–274°F.

origins
The origin of the moon is still open to debate. Scientists suggest the following theories: that it split from Earth; that it was a separate body captured by Earth's

gravity; that it formed in orbit around Earth; or that it was formed from debris thrown off when a body the size of Mars struck Earth.

research

The far side of the moon was first photographed from the Soviet *Lunik 3* probe in October 1959. Much of our information about the moon has been derived from this and other photographs and measurements taken by U.S. and Soviet moon probes, from geological samples brought back by U.S. *Apollo* astronauts and by Soviet *Luna* probes, and from experiments set up by the U.S. astronauts (1969–1972). The U.S. probe *Lunar Prospector* was launched in January 1998 to examine the composition of the lunar crust, record gamma rays, and map the lunar magnetic field. It was the *Lunar Prospector* that discovered the ice on the moon, in March 1998.

composition

The moon's composition is rocky, with a surface heavily scarred by meteorite impacts that have formed craters up to 240 km/150 mi across. Seismic observations indicate that the moon's surface extends downward for tens of kilometers; below this crust is a solid mantle about 1,100 km/688 mi thick; and below that, a silicate core, part of which may be molten. Rocks brought back by astronauts show the moon is 4.6 billion years old, the same age as Earth. It is made up of the same chemical elements as Earth, but in different proportions, and differs from Earth in that most of the moon's surface features were formed within the first billion years of its history, when it was hit repeatedly by meteorites.

The youngest craters are surrounded by bright rays of ejected rock. The largest scars have been filled by dark lava to produce the lowland plains called seas, or maria (plural of mare). These dark patches form the so-called "man-in-the-moon" pattern. Inside some craters that are permanently in shadow is up to 300 million metric tons/330 million tons of ice existing as a thin layer of crystals. One of the moon's easiest features to observe is the Mare Plato, which is about 100 km/62 mi in diameter and 2,700 m/8,860 ft deep and, at times, is visible with the naked eye alone.

The U.S. lunar probe *Clementine* discovered an enormous crater on the far side of the moon in 1994. The South Pole–Aitken crater is 2,500 km/1,563 mi across and 13 km/8 mi deep, making it the largest-known crater in the solar system.

moon probes

Crewless spacecraft used to investigate the moon. Early probes flew past the moon or crash-landed on it, but later ones achieved soft landings or went into orbit. Soviet probes included the *Luna/Lunik* series. U.S. probes *(Ranger, Surveyor, Lunar Orbiter)* prepared the way for the *Apollo* crewed flights.

The first space probe to hit the moon was the Soviet *Luna 2*, on September 13, 1959 (*Luna 1* had missed the moon eight months earlier). In October 1959, *Luna*

3 sent back the first photographs of the moon's far side. *Luna 9* was the first probe to soft-land on the moon, on February 3, 1966, transmitting photographs of the surface to Earth. In September 1970, *Luna 16* was the first probe to return automatically to Earth carrying moon samples, although by then moon rocks had already been brought back by U.S. *Apollo* astronauts. *Luna 17* landed in November 1970 carrying a lunar rover, *Lunokhod,* which was driven over the moon's surface by remote control from Earth.

The first successful U.S. moon probe was *Ranger 7,* which took close-up photographs before it hit the moon on July 31, 1964. *Surveyor 1,* on June 2, 1966, was the first U.S. probe to soft-land on the lunar surface. It took photographs, and later *Surveyors* analyzed the surface rocks. Between 1966 and 1967, a series of five U.S. *Lunar Orbiters* photographed the entire moon in detail, in preparation for the *Apollo* landings (1969–1972).

In March 1990, Japan put a satellite, *Hagoromo,* into orbit around the moon. Weighing only 11 kg/26 lb, it was released from a larger Japanese space probe.

Mount Palomar
Astronomical observatory, near Pasadena, California. Its telescopes include the 5-m/200-in Hale Telescope and the 1.2-m/48-in Schmidt Telescope. Completed in 1948, it was the world's premier observatory during the 1950s.

Mount Stromlo Observatory
Astronomical observatory established in Canberra, Australia, in 1923. Important observations have been made there of the Magellanic Clouds, which can be seen clearly from southern Australia.

Mount Wilson
Site, near Los Angeles, California, of the 2.5-m/100-in Hooker Telescope, opened in 1917, with which U.S. astronomer Edwin Hubble discovered the expansion of the universe. Two solar telescopes in towers 18.3 m/60 ft and 45.7 m/150 ft tall, and a 1.5- m/60-in reflector opened in 1908, also operate there.

Mullard Radio Astronomy Observatory
Radio observatory of Cambridge University, England. Its main instrument is the Ryle Telescope, eight dishes 12.8-m/42-ft wide, four are mounted on a rail track, and others are fixed at 1.2-km/0.7-mi intervals, in a line of 5-km/3 mi-long, opened in 1972.

Multiple Mirror Telescope
Telescope on Mount Hopkins, Arizona, opened in 1979, consisting of six 1.83-m/72-in mirrors mounted in a hexagon, the light-collecting area of which equals that of a single mirror of 4.5-m/176-in diameter. The six mirrors were replaced in 1996 with a single mirror 6.5 m/21.3 ft wide.

Musca

Small constellation of the Southern Hemisphere, represented as a fly. It was some-times known as Musca Australis to distinguish it from Musca Borealis, a former constellation close to Aries.

nadir

The point on the celestial sphere vertically below the observer and, hence, dia-metrically opposite the zenith. The term is used metaphorically to mean the low point of a person's fortunes.

National Aeronautics and Space Administration (NASA)

U.S. government agency for spaceflight and aeronautical research, founded in 1958 by the National Aeronautics and Space Act. Its headquarters are in Washington, D.C., and its main installation is at the Kennedy Space Center in Florida. NASA's early planetary and lunar programs included *Pioneer* spacecraft, from 1958, which gathered data for the later crewed missions, the most famous of which took the first people to the moon in *Apollo 11* (July 16–24, 1969).

In the early 1990s, NASA moved toward lower-budget missions, called *Discovery*, which should not exceed a budget of about $200 million (excluding launch costs) nor a development period of three years. The first missions were the *Near Earth Asteroid Rendezvous* craft and the *Lunar Prospector*.

nebula

Cloud of gas and dust in space. Nebulae are the birthplaces of stars, but some nebulae are produced by gas thrown off from dying stars. Nebulae are classified depending on whether they emit, reflect, or absorb light.

An emission nebula, such as the Orion Nebula, glows brightly because its gas is energized by stars that have formed within it. In a reflection nebula, starlight reflects off grains of dust in the nebula, such as surround the stars of the Pleiades cluster. A dark nebula is a dense cloud, composed of molecular hydrogen, which partly or completely absorbs light behind it. Examples include the Coalsack Nebula in Crux and the Horsehead Nebula in Orion.

nebular hypothesis

Hypothesis that the solar system evolved from a nebula. Eighteenth-century Swed-ish mystic and scientist Emanuel Swedenborg and the eighteenth-century Ger-man philosopher Immanuel Kant both put forward explanations of this kind, but the nineteenth-century French astronomer Pierre Laplace was the first to develop a nebular hypothesis on strictly scientific lines. Laplace's hypothesis was that ro-tating nebulae were first formed by the condensation of gaseous matter and that the sun was originally a nebula of this type. As the nebula continued to shrink, it rotated faster and faster, until fragments broke away from the main body of the sun. These fragments condensed under their own gravitational attraction and formed the planets. Laplace's hypothesis is untenable, as it fails to explain the fact that nearly all of the angular momentum of the solar system resides in the planets.

The tidal hypothesis of the early-twentieth-century British mathematician and astrophysicist James Jeans, that the planets were formed from material torn out of the sun by the close approach of a passing star, is also untenable, as the ejected material would disperse into space. Modern theories offer no unique solution but agree that the sun and the planets condensed simultaneously from a rotating cloud of gas and dust.

nemesis theory

Theory of animal extinction, suggesting that a sister star to the sun caused the extinction of the dinosaurs and other groups of animals. The theory holds that the movement of this as yet undiscovered star disrupts the Oort cloud of comets every twenty-six million years, resulting in Earth suffering an increased bombardment from comets at these times. The theory was proposed in 1984 to explain the newly discovered layer of iridium—an element found in comets and meteorites—in rocks dating from the end of dinosaur times. However, many paleontologists deny that there is any evidence for a twenty-six-million-year cycle of extinctions.

Neptune

The eighth planet in average distance from the sun. It is a giant gas (hydrogen, helium, methane) planet, with a mass 17.2 times that of Earth. It has the highest winds in the solar system.

mean distance from the sun:
4.4 billion km/2.794 billion mi

equatorial diameter:
48,600 km/30,200 mi

rotation period:
0.67 Earth day

year:
164.8 Earth years

atmosphere:
Consists primarily of hydrogen (85 percent) with helium (13 percent) and methane (1–2 percent). Methane in Neptune's atmosphere absorbs red light and gives the planet a blue coloring.

surface:
Hydrogen, helium, and methane. Its interior is believed to consist of a central rocky core covered by a layer of ice.

satellites:
Of Neptune's eight moons, two (Triton and Nereid) are visible from Earth. Six were discovered by the U.S. *Voyager 2* probe in 1989, of which Proteus (diameter 415 km/260 mi) is larger than Nereid (300 km/200 mi).

rings:
There are four faint rings: Galle, Leverrier, Arago, and Adams (in order from Neptune). Galle is the widest, at 1,700 km/1,060 mi. Leverrier and Arago are divided by a wide diffuse particle band called the plateau.

Neptune was located in 1846 by German astronomers Johan Galle and Heinrich d'Arrest after calculations by English astronomer John Couch Adams and French mathematician Urbain Leverrier had predicted its existence from disturbances in the movement of Uranus. *Voyager 2,* which passed Neptune in August 1989, revealed various cloud features, notably an Earth-size oval storm cloud, the Great Dark Spot, similar to the Great Red Spot on Jupiter, but images taken by the Hubble Space Telescope in 1994 show that the Great Dark Spot has disappeared. A smaller dark spot, DS2, has also gone.

neutron star
Very small, "superdense" star composed mostly of neutrons. They are thought to form when massive stars explode as supernovae, during which the protons and the electrons of the star's atoms merge, owing to intense gravitational collapse, to make neutrons. A neutron star has the mass of two to three suns, compressed into a globe only 20 km/12 mi in diameter. If its mass is any greater, its gravity will be so strong that it will shrink even further to become a black hole. Being so small, neutron stars can spin very quickly. The rapidly flashing radio stars called pulsars are believed to be neutron stars. The flashing is caused by a rotating beam of radio energy similar in behavior to a lighthouse beam of light.

New General Catalog
Catalog of star clusters and nebulae compiled by Danish astronomer John Louis Emil Dreyer and published in 1888. Its main aim was to revise, correct, and expand upon the *General Catalogue* compiled by English astronomer John Herschel, which appeared in 1864.

New Technology Telescope
Optical telescope that forms part of the European Southern Observatory, La Silla, Chile; it came into operation in 1990. It has a thin, lightweight mirror 3.5 m/138 in across, which is kept in shape by computer-adjustable supports to produce a sharper image than is possible with conventional mirrors. Such a system is termed active optics.

Newtonian telescope
A simple reflecting telescope in which light collected by a parabolic primary mirror is directed to a focus at the side of the tube by a flat secondary mirror placed

at 45° to the optical axis. It is named for British physicist and mathematician Isaac Newton, who constructed such a telescope in 1668.

Norma

Small constellation of the Southern Hemisphere, formerly known as Norma et Regula, "the Level and Square." It was named by eighteenth-century French scientist and astronomer Nicolas Lacaille at the same time as the adjacent Circinus, represented as a pair of compasses.

North Star

See Polaris, or Pole Star, or North Star.

northern lights

See aurora.

nova (plural novae)

Faint star that suddenly erupts in brightness by 10,000 times or more, remains bright for a few days, and then fades away and is not seen again for very many years, if at all. Novae are believed to occur in close binary-star systems, where gas from one star flows to a companion white dwarf. The gas ignites and is thrown off in an explosion at speeds of 1,500 kps/930 mps or more. Unlike a supernova, the star is not completely disrupted by the outburst. After a few weeks or months, it subsides to its previous state; it may erupt many more times.

Although the name comes from the Latin for "new," photographic records show that such stars are not really new but are faint stars undergoing an outburst of radiation that temporarily gives them an absolute magnitude in the range of 6–10, at least 100,000 times brighter than the sun. They fade away, rapidly at first and then more slowly over several years. Two or three such stars are detected in our galaxy each year, but, on average, one is sufficiently close to us to become a conspicuous naked-eye object only about once in ten years. Novae similar to those appearing in our own galaxy have also been observed in other galaxies.

named novae

Novae are named according to the constellation and the year in which they appear. Bright ones in the twentieth century that have been intensively studied spectroscopically include Nova Persei 1901, Nova Geminorum 1912, Nova Aquilae 1918 (which became almost as bright as Sirius), Nova Pictoris 1925, Nova Herculis 1934, Nova Puppis 1942, and Nova Cygni 1975. Nova Cygni was first seen on August 29, 1975, with a magnitude of 3.0. By August 31, it had attained its maximum brightness of 1.8, which it retained for a day, and it was below naked-eye visibility by September 5. It is visible on photographs taken at the Riga Observatory on August 5, 8, and 12, 1975, only a few days before the outburst, as a star having a yellow magnitude of 16.

the phenomenon
The rate of increase in brightness, the absolute magnitude, and the rate of fading differ from nova to nova, but spectroscopic observations indicate that the phenomenon in each case is essentially the same. The star blows off its outer shell with velocities of ejection of the order of 1,500 kps/930 mps, the total mass of the shell being about that of the sun. The growing shell initially behaves like the photosphere of a supergiant and pours out rapidly increasing amounts of white light. As the expansion continues, however, it grows more diffuse, so that conditions in it become more like those in a gaseous nebula. The continuous spectrum fades, leaving mainly monochromatic radiations that decay much more slowly.

theory
Many novae have been found to be members of close binary systems, and, consequently, it has been hypothesized that the stars subject to a nova outburst are close binaries that have evolved so far that the initially larger member has already become a white dwarf, while the other one is in its red-giant phase. As the outer layers of the red giant swell out beyond the Lagrangian surface, some of the material is attracted to the white dwarf, where the surface gravity is so high that the extra matter produces a sufficient rise in the pressure and temperature to start proton-proton reactions in the hydrogen still remaining in the outer layers. The energy thus suddenly released ejects the surface layer into space, so that the remainder can relapse to its former white-dwarf state. This theory implies that the same system may suffer several nova outbursts, and this does seem to be the case: There are a number of stars, the so-called recurrent novae, for which more than one such outburst has been observed.

nucleus
The compact central core of a galaxy, often containing powerful radio, X-ray, and infrared sources. Active galaxies have extremely energetic nuclei.

nutation
A slight "nodding" of Earth in space, caused by the varying gravitational pulls of the sun and the moon. Nutation changes the angle of Earth's axial tilt (average 23.5°) by about 9 seconds of arc to either side of its mean position, a complete cycle taking just more than 18.5 years.

observatory
Site or facility for observing astronomical or meteorological phenomena. The earliest recorded observatory was in Alexandria, Egypt, built by Ptolemy Soter in about A.D. 100. The modern observatory dates from the invention of the telescope. Observatories may be ground based, carried on aircraft, or sent into orbit as satellites, in space stations, and on the space shuttle. Most early observatories were

near towns, but, with the advent of big telescopes, clear skies with little background light and, hence, high, remote sites became essential.

The most powerful optical telescopes covering the sky are at Mauna Kea, Hawaii; Mount Palomar, California; Kitt Peak National Observatory, Arizona; La Palma, Canary Islands; Cerro Tololo Interamerican Observatory and the European Southern Observatory, both in Chile; Siding Spring Mountain, Australia; and Zelenchukskaya, Russia.

Radio astronomy observatories include Jodrell Bank, Cheshire, England; the Mullard Radio Astronomy Observatory, Cambridge, England; Arecibo, Puerto Rico; Effelsberg, Germany; and Parkes, Australia. The Hubble Space Telescope was launched into orbit in 1990. The Very Large Telescope, constructed by the European Southern Observatory in the mountains of northern Chile at Cerro Paranal, transmitted its first images in 1997.

occultation

The temporary obscuring of a star by a body in the solar system. Occultations are used to provide information about changes in an orbit and the structure of objects in space, such as radio sources. The exact shapes and sizes of planets and asteroids can be found when they occult stars. The rings of Uranus were discovered when that planet occulted a star in 1977.

Octans

Faint constellation in the Southern Hemisphere, represented as an octant. It contains the southern celestial pole. The closest naked-eye star to the south celestial pole is fifth-magnitude Sigma Octantis.

Olbers's paradox

Question put forward 1826 by the German doctor, mathematician, and astronomer Heinrich Olbers, who asked: If the universe is infinite in extent and filled with stars, why is the sky dark at night? The answer is that the stars do not live infinitely long, so there is not enough starlight to fill the universe. A wrong answer, frequently given, is that the expansion of the universe weakens the starlight.

omega

Last letter of the Greek alphabet ($\grave{\mathrm{U}}$), used as a symbol for the mass density of the universe. If $\grave{\mathrm{U}}$ is less than 1.0, the universe will expand forever; if it is more than 1.0, the gravitational pull of its mass will be strong enough to reverse its expansion and cause its eventual collapse. The value of $\grave{\mathrm{U}}$ is estimated as being between 0.1 and 1.0.

Oort cloud

Spherical cloud of comets beyond Pluto, extending out to about 100,000 astronomical units (1.0 light-year) from the sun. The gravitational effect of passing stars and the rest of our galaxy disturbs comets from the cloud, so that they fall in toward the sun on highly elongated orbits, becoming visible from Earth. As many as ten trillion comets may reside in the Oort cloud, named for Dutch astronomer Jan Hendrik Oort, who postulated it in 1950.

open cluster, or galactic cluster
A loose cluster of young stars. More than twelve hundred open clusters have been cataloged, each containing between a dozen and several thousand stars. They are of interest to astronomers because they represent samples of stars that have been formed at the same time from similar material. Examples include the Pleiades and the Hyades.

Ophiuchus
Large constellation along the celestial equator, known as the Serpent Bearer because the constellation Serpens is wrapped around it. The sun passes through Ophiuchus each December, but the constellation is not part of the zodiac. Ophiuchus contains Barnard's Star.

opposition
The moment at which a body in the solar system lies opposite the sun in the sky as seen from Earth and crosses the Meridian at about midnight. Although the inferior planets cannot come to opposition, it is the best time for observation of the superior planets, since they can then be seen all night.

orbit
Path of one body in space around another, such as the orbit of Earth around the sun or the moon around Earth. When the two bodies are similar in mass, as in a binary star, both bodies move around their common center of mass. The movement of objects in orbit follows Kepler's laws, which apply to artificial satellites as well as to natural bodies.

As stated by the laws, formulated by German mathematician and astronomer Johannes Kepler in the seventeenth century, the orbit of one body around another is an ellipse. The ellipse can be highly elongated, as are comet orbits around the sun, or it may be almost circular, as are those of some planets. The closest point of a planet's orbit to the sun is called perihelion; the most distant point is aphelion. (For a body orbiting Earth, the closest and farthest points of the orbit are called perigee and apogee, respectively.)

Orion
Prominent constellation in the equatorial region of the sky, identified with the hunter of Greek mythology.

The bright stars Alpha (Betelgeuse), Gamma (Bellatrix), Beta (Rigel), and Kappa Orionis mark the shoulders and legs of Orion. Between them the belt is formed by Delta, Epsilon, and Zeta, three second-magnitude stars equally spaced in a straight line. Beneath the belt is a line of fainter stars marking Orion's sword. One of these, Theta, is not really a star but the brightest part of the Orion Nebula. Nearby is one of the most distinctive dark nebulae, the Horsehead.

Both Betelgeuse and Rigel are first-magnitude stars and are supergiants, but, while Betelgeuse is red, Rigel, like most of the other bright stars in the Orion area, is blue-white.

The Orionid meteors, which appear October 18–26, are associated with Halley's comet.

Orion Nebula
Luminous cloud of gas and dust 1,500 light-years away, in the constellation Orion, from which stars are forming. It is about 15 light-years in diameter and contains enough gas to make a cluster of thousands of stars. At the nebula's center is a group of hot young stars called the Trapezium, which make the surrounding gas glow. The nebula is visible to the naked eye as a misty patch below the belt of Orion.

oscillating universe
A theory that states that the gravitational attraction of the mass within the universe will eventually slow down and stop the expansion of the universe. The outward motions of the galaxies will then be reversed, eventually resulting in a Big Crunch, in which all of the matter in the universe would be contracted into a small volume of high density. This could undergo a further Big Bang, thereby creating another expansion phase. The theory suggests that the universe would alternately expand and collapse through alternate Big Bangs and Big Crunches.

outer space
See space, or outer space.

parallax
The change in the apparent position of an object against its background when viewed from two different positions. In astronomy, nearby stars show a shift owing to parallax when viewed from different positions on Earth's orbit around the sun. A star's parallax is used to deduce its distance from Earth. Nearer bodies such as the moon, sun, and planets also show a parallax caused by the motion of Earth. Diurnal parallax is caused by Earth's rotation.

Parkes Observatory
Site in New South Wales, part of the Australia Telescope National Facility, featuring a radio telescope of 64-m/210-ft aperture, run by the Commonwealth Scientific and Industrial Research Organization (CSIRO). It received a U.S. National Aeronautics and Space Administration (NASA)-funded upgrade in 1996 to enable it to track the space probe *Galileo*.

parsec
A unit (symbol pc) used for distances to stars and galaxies. One parsec is equal to 3.2616 light-years, 2.063×10^5 astronomical units, and 3.0857×10^{12} km. It is the

distance at which a star would have a parallax (apparent shift in position) of one second of arc when viewed from two points the same distance apart as Earth's distance from the sun; or the distance at which one astronomical unit subtends an angle of one second of arc.

Pavo
Constellation of the Southern Hemisphere, represented as a peacock.

Pegasus
Constellation of the Northern Hemisphere, near Cygnus, represented as the winged horse of Greek mythology. It is the seventh-largest constellation in the sky, and its main feature is a square outlined by four stars, one of which (Alpherat) is actually part of the adjoining constellation Andromeda. Diagonally across is Markab (or Alpha Pegasi), about 100 light-years distant.

periastron
The point at which an object traveling in an elliptical orbit around a star is at its closest to the star; the point at which it is farthest is known as the apastron.

perigee
The point at which an object, traveling in an elliptical orbit around Earth, is at its closest to Earth. The point at which it is farthest from Earth is the apogee.

perihelion
The point at which an object, traveling in an elliptical orbit around the sun, is at its closest to the sun. The point at which it is farthest from the sun is the aphelion.

Perseus
Bright constellation of the Northern Hemisphere, near Cassiopeia. It is represented as the mythological hero; the head of the decapitated Gorgon, Medusa, is marked by Algol (Beta Persei), the best-known of the eclipsing binary stars. Perseus lies in the Milky Way and contains the Double Cluster, a twin cluster of stars called h and Chi Persei. They are just visible to the naked eye as two hazy patches of light close to each other. Every August, the Perseid meteor shower radiates from the constellation's northern part. The meteor shower is a remnant of Comet 1862 III.

phase
See moon (Earth's).

Phobos
One of the two moons of Mars. It is an irregularly shaped lump of rock, cratered by meteorite impacts. Phobos is 27 x 22 x 19 km/17 x 13 x 12 mi across and orbits Mars every 0.32 days at a distance of 9,400 km/5,840 mi from the planet's center. It is thought to be an asteroid captured by Mars's gravity.

Phoenix
Constellation of the Southern Hemisphere.

Pholus, or 5145 Pholus
A red centaur (star) discovered in 1991.

photosphere
Visible surface of the sun, which emits light and heat. About 300 km/200 mi deep, it consists of incandescent gas at a temperature of 5,800 K (5,530°C/9,986°F). Rising cells of hot gas produce a mottling of the photosphere known as granulation, each granule being about 1,000 km/620 mi in diameter. The photosphere is often marked by large, dark patches called sunspots.

Pictor
Constellation of the Southern Hemisphere, originally named Equuleus Pictoris, "the Painter's Easel," by eighteenth-century French astronomer Nicolas Lacaille. Its name now refers only to the painter.

Pioneer probe
Any of a series of U.S. solar-system space probes (1958–1978). *Pioneer 1–3,* all launched in 1958, were intended moon probes, but *Pioneer 2's* launch failed, and *1* and *3* failed to reach their target, although they did measure the Van Allen radiation belts. *Pioneer 4* began to orbit the sun after passing the moon.

Pioneer 5 (1960) was the first of a series to study the solar wind between the planets. *Pioneer 6* (1965), *7* (1966), *8* (1967), and *9* (1968) went into orbit around the sun and monitored solar activity. *Pioneer 10* (March 1972) was the first probe to reach Jupiter (December 1973) and to leave the solar system (1983). *Pioneer 11* (April 1973) passed Jupiter (December 1974) and was the first probe to reach Saturn (September 1979) before also leaving the solar system. The U.S. National Aeronautics and Space Administration (NASA) ceased to operate *Pioneer 10* in April 1997. The probe had functioned for twenty-five years and reached a distance of 10 billion km/6.25 billion mi from the sun. *Pioneer 11* ceased to function in 1995. *Pioneer 10* and *11* carry plaques containing messages from Earth in case they are found by other civilizations among the stars.

Pioneer-Venus probes were launched in May and August 1978. One orbited Venus, and the other dropped three probes onto the surface. The orbiter finally burned up in the atmosphere of Venus in 1992.

Pisces
Inconspicuous zodiac constellation, mainly in the Northern Hemisphere between Aries and Aquarius, near Pegasus. It is represented as two fish tied together by their tails. The Circlet, a delicate ring of stars, marks the head of the western fish in Pisces. The constellation contains the vernal equinox, the point at which the sun's path around the sky (the ecliptic) crosses the celestial equator. The sun reaches this point around March 21 each year as it passes through Pisces from

mid-March to late April. In astrology, the dates for Pisces are about February 19–March 20.

Piscis Austrinus
Constellation of the Southern Hemisphere near Capricornus, popularly known as Southern Fish. Its brightest star is the first-magnitude Fomalhaut.

Pistol star
Recognized as the most massive known star by a team at the University of California at Los Angeles (UCLA) led by U.S. astronomer Don Figer. It emits ten million times the luminosity of the sun and is 25,000 light-years from Earth.

plage
Bright patch in the chromosphere above a group of sunspots, occasionally seen on images of the sun in hydrogen light.

planet
From the Greek for "wanderer," a large celestial body in orbit around a star, composed of rock, metal, or gas. There are nine planets in the solar system: Mercury, Venus, Earth, Mars, Jupiter, Saturn, Neptune, Uranus, and Pluto. The inner four, called the terrestrial planets, are small and rocky, and include the planet Earth. The outer planets, with the exception of Pluto, are called the major planets and consist of large balls of rock, liquid, and gas; the largest is Jupiter, which contains a mass equivalent to 70 percent of all of the other planets combined. Planets do not produce light but reflect the light of their parent star.

As seen from Earth, all of the historic planets are conspicuous naked-eye objects moving in looped paths against the stellar background. The size of each loop, which is caused by Earth's own motion around the sun, is inversely proportional to the planet's distance from Earth.

In 1995, Italian astronomers believed that they had detected a new planet around 51 Pegasi in the constellation Pegasus. It was named 51 Pegasi B and is thought to have a mass comparable to that of Jupiter. The gravitational pull thought to be that of the planet may be caused by pulsation in the parent star.

The discovery of three further new planets was announced at the American Astronomical Society meeting in January 1996. All are outside the solar system, but two are only about 35 light-years from Earth and orbit stars visible to the naked eye. One, 70 Vir B, is in the constellation Virgo, and the other, 47 UMa B, is in Ursa Major. The third, ß Pictoris, is about 50 light-years away in the southern constellation Pictor.

In April 1996, another planet was discovered, orbiting Rho Cancri in the constellation Cancer. Yet another was found in June 1996, this one orbiting the star Tau Boötis. By July 1996, the total of new planets discovered since October 1995 had risen to ten.

In October 1996, U.S. astronomers announced the discovery of a new planet with an orbit that is more irregular than that of any other. The new planet has 1.6

times the mass of Jupiter and orbits the star 16 Cygni B. Its distance from 16 Cygni B varies from 90 million km/55.9 million mi to 390 million km/242 million mi.

planetarium
Optical projection device by means of which the motions of stars and planets are reproduced on a domed ceiling representing the sky. The planetarium of the Heureka Science Center, Finland, opened in 1989, was the world's first to use fiber optics.

planetary embryo
One of numerous massive bodies thought to have formed from the accretion of planetesimals during the formation of the solar system. Embryos in the region of Earth's orbit would have been about 10^{23} kg in mass, and about ten to one hundred of them would have coalesced to make Earth.

planetary nebula
Shell of gas thrown off by a star at the end of its life. Planetary nebulae have nothing to do with planets. They were named by German-born British astronomer William Herschel, who thought that their rounded shape resembled the disk of a planet. After a star such as the sun has expanded to become a red giant, its outer layers are ejected into space to form a planetary nebula, leaving the core as a white dwarf at the center.

planetesimal
Body of rock in space, smaller than a planet, attracted to other such bodies during planet formation. According to modern solar-nebula theory, the sun and the planets are thought to have formed from a rotating dust cloud generated by a supernova explosion. On condensation, this cloud formed a central sun and a rotating disk, the material of which separated into rings of dust grains that began to stick together. Larger and larger clumps formed in each ring and, eventually, collected into bodies the size of present-day asteroids, called planetesimals. After numerous collisions, these bodies eventually formed the nucleus of the various planets of our solar system.

A once-popular theory, called the planetesimal theory, held that planetesimals were formed from giant tongues of solar material torn from a preexisting sun by the gravitational attraction of a passing star. These planetesimals then ended up in orbits around the leftover sun in the plane of the passing star. As in the modern theory, the planetesimals then collided and coalesced into the present planets.

planisphere
A graphical device for determining the aspect of the sky for any date and time in the year. It consists of two disks mounted concentrically, so that the upper disk, which has an aperture corresponding to the horizon of the observer, can rotate over the lower disk, which is printed with a map of the sky centered on the north or south celestial pole. In use, the observer aligns the time of day marked around

the edge of the upper disk with the date marked around the edge of the lower disk. The aperture then shows which stars are above the horizon.

Plaskett's Star
The most massive binary star known, consisting of two supergiants of about forty to fifty solar masses, orbiting each other every 14.4 days. Plaskett's Star lies in the constellation Monoceros, named for the unicorn.

Pleiades
An open star cluster about 400 light-years away from Earth in the constellation Taurus, represented as the Seven Sisters of Greek mythology. Its brightest stars (highly luminous blue-white giants only a few million years old) are visible to the naked eye, but there are many fainter ones. It is a young cluster, and the stars of the Pleiades are still surrounded by traces of the reflection nebula from which they formed, visible on long-exposure photographs.

The cluster contains about a dozen blue stars visible to the naked eye, spread over an area about twice that of the full moon. The six brightest stars are easily seen; the seventh is more difficult to see and is supposed to represent the lost sister, Electra. It takes a clear sky to distinguish the next four or five. Binoculars show at least fifty stars and a telescope hundreds more.

According to Greek mythology, the Pleiades were the seven daughters of Atlas and Pleione; the eldest, Electra, was "lost" because she married a mortal. Nine of the stars are named for the Seven Sisters and their parents: Alcyone, Maia, Electra, Merope, Taygeta, Celaeno, Asterope, Atlas, and Pleione.

Plesetsk
Rocket-launching site 170 km/105 mi south of Archangel, Russia. Since 1966, the USSR launches artificial satellites, mostly military ones, from there.

Plow, the
Another name for the Big Dipper, the most prominent part of the constellation Ursa Major.

Pluto
The smallest and, usually, the outermost planet of the solar system. The existence of Pluto was predicted by calculation by U.S. astronomer Percival Lowell, and the planet was located by U.S. astronomer Clyde Tombaugh in 1930. Its highly elliptical orbit occasionally takes it within the orbit of Neptune, as in the period 1979–1999. Pluto has a mass about 0.002 of that of Earth.

mean distance from the sun:
5.8 billion km/3.6 billion mi

equatorial diameter:
2,300 km/1,438 mi

rotation period:
6.39 Earth days

year:
248.5 Earth years

atmosphere:
Thin atmosphere with small amounts of methane gas

surface:
Low density, composed of rock and ice, primarily frozen methane; there is an ice cap at Pluto's north pole. The Hubble Space Telescope photographed Pluto's surface in 1996.

satellites:
One moon, Charon, discovered in 1978 by U.S. astronomer James Christy. It is about 1,200 km/750 mi in diameter, half the size of Pluto, making it the largest moon in relation to its parent planet in the solar system. It orbits about 20,000 km/12,500 mi from the planet's center every 6.39 days—the same time that Pluto takes to spin on its axis. Charon is composed mainly of ice. Some astronomers have suggested that Pluto was a former moon of Neptune that escaped, but it is more likely that it was an independent body that was captured by the sun's gravitational field.

polarimetry
Any technique for measuring the degree of polarization of radiation from stars, galaxies, and other objects.

Polaris, or Pole Star, or North Star
Bright star closest to the north celestial pole and the brightest star in the constellation Ursa Minor. Its position is indicated by the "pointers" in Ursa Major. Polaris is a yellow supergiant about 500 light-years away from Earth. It is also known as Alpha Ursae Minoris.

It lies within 1° of the north celestial pole; precession (Earth's axial wobble) will bring Polaris closest to the celestial pole (less than 0.5° away) in about A.D. 2100. Then its distance will start to increase, reaching 1° in 2205 and 47° in 2800. Other bright stars that have been, or will be, close to the north celestial pole are Alpha Draconis (2800 B.C.), Gamma Cephei (A.D. 4000), Alpha Cephei (A.D. 7000), and Vega (A.D. 14000).

Polaris is a Cepheid variable whose magnitude varies between 2.1 and 2.2 over 3.97 days.

Pole Star
See Polaris, or Pole Star, or North Star.

Pollux, or Beta Geminorum

Brightest star in the constellation Gemini, and the seventeenth-brightest star in the sky. Pollux is a yellow star with a true luminosity forty-five times that of the sun. It is 36 light-years away from Earth. The first-magnitude Pollux and the second-magnitude Castor (Alpha Geminorum) mark the heads of the Gemini twins. It is thought that the two stars may have changed their relative brightness since Bayer named them, as the word "Alpha" is usually assigned to the brightest star in a constellation.

Praesepe, or the Beehive

One of the nearer open star clusters, located in the constellation of Cancer. To the naked eye, it appears as a hazy patch, but a small telescope shows it to be a cluster of separate stars. It is very similar to the Hyades cluster but is more than four times farther from Earth.

precession

Slow wobble of Earth on its axis, like that of a spinning top. The gravitational pulls of the sun and the moon on Earth's equatorial bulge cause Earth's axis to trace out a circle on the sky every 25,800 years. The position of the celestial poles is constantly changing owing to precession, as are the positions of the equinoxes (the points at which the celestial equator intersects the sun's path around the sky). The precession of the equinoxes refers to a gradual westward drift in the ecliptic (the path that the sun appears to follow) and in the coordinates of objects on the celestial sphere.

This is why the dates of the astrological signs of the zodiac no longer correspond to the times of year when the sun actually passes through the constellations. For example, the sun passes through Leo from mid-August to mid-September, but the astrological dates for Leo are between about July 23 and August 22.

Precession also occurs in other planets. Uranus has the solar system's fastest-known precession (264 days), determined in 1995.

primary wave

See P-wave, or primary wave.

Procyon, or Alpha Canis Minoris

Brightest star in the constellation Canis Minor and the eighth-brightest star in the sky. Procyon is a first-magnitude white star 11.4 light-years from Earth, with a mass of 1.7 suns. It has a white-dwarf companion that orbits it every forty years.

The name, derived from Greek, means "before the dog" and reflects the fact that, in mid-northern latitudes, Procyon rises shortly before Sirius, the Dog Star. Procyon and Sirius are sometimes called "the Dog Stars." Both are relatively close to us and have white-dwarf companions.

prograde, or direct motion
In astronomy, the orbit or rotation of a planet, or of a satellite if the sense of rotation is the same as the general sense of rotation of the solar system. On the celestial sphere, it refers to motion from west to east against the background of stars.

prominence
Bright cloud of gas projecting from the sun into space 100,000 km/60,000 mi or more. Quiescent prominences last for months and are held in place by magnetic fields in the sun's corona. Surge prominences shoot gas into space at speeds of 1,000 kps/600 mps. Loop prominences are gases falling back to the sun's surface after a solar flare.

proper motion
Gradual change in the position of a star that results from its motion in orbit around our galaxy, the Milky Way. Proper motions are slight and undetectable to the naked eye but can be accurately measured on telescopic photographs taken many years apart. Barnard's Star has the largest proper motion, 10.3 arc seconds per year.

Proton rocket
Soviet space rocket introduced in 1965, used to launch heavy satellites, space probes, and the *Salyut* and *Mir* space stations. *Proton* consists of up to four stages as necessary. It has never been used to launch humans into space.

protostar
Early formation of a star that has recently condensed out of an interstellar cloud and that is not yet hot enough for hydrogen burning to start. Protostars derive their energy from gravitational contraction.

Proxima Centauri
The closest star to the sun, 4.2 light-years away from Earth. It is a faint red dwarf, visible only with a telescope, and is a member of the Alpha Centauri triple-star system. It is called Proxima because it is 0.1 light-years closer to us than its two partners.

pulsar
Celestial source that emits pulses of energy at regular intervals, ranging from a few seconds to a few thousandths of a second. Pulsars are thought to be rapidly rotating neutron stars, which flash at radio and other wavelengths as they spin. They were first discovered in 1967 by British astronomers Jocelyn Bell and Antony Hewish at the Mullard Radio Astronomy Observatory, Cambridge, England. More than five hundred radio pulsars are known in our galaxy, although a million or so may exist.

Pulsars slow down as they get older, and, eventually, the flashes fade. Of the known radio pulsars, twenty are millisecond pulsars (flashing a thousand times a second). Such pulsars are thought to be more than a billion years old. Two pulsars, one (estimated to be a thousand years old) in the Crab Nebula and one (estimated to be eleven thousand years old) in the constellation Vela, give out flashes of visible light.

Pulsars were first identified as compact radio sources regularly emitting very rapid, intense pulses of radiation. Within a year of their discovery, though, they were identified with rapidly rotating neutron stars formed by the gravitational collapse of stars in supernovae explosions.

Pulsars have proved useful for investigating the properties of the ionized gas that permeates interstellar space. It has also been possible to use these properties to estimate the distances of the pulsars and their velocities through space. Most of the pulsars so far observed are close to the plane of the galaxy and are more numerous in the direction of its center. Their velocities are relatively high. It may be that pulsars are forcibly ejected during the supernovae explosions in which they are formed and that their speed is the reason that they are now found close to only the youngest supernovae remnants.

The closest pulsar to Earth, PSR J0108–1431, lies 280 light-years away in the constellation of Cetus; it was discovered in 1994.

Puppis
Bright constellation of the Southern Hemisphere, representing a ship's poop or stern. It was originally regarded as part of Argo.

P-wave, or primary wave
In seismology, a class of seismic wave that passes through Earth in the form of longitudinal pressure waves at speeds of 6–7 kps/3.7–4.4 mps in the crust and up to 13 kps/8 mps in deeper layers. P-waves from an earthquake travel faster than S-waves (secondary waves) and are the first to arrive at monitoring stations (hence, primary waves). They can travel through both solid rock and the liquid outer core of Earth.

Pyxis
Small constellation of the Southern Hemisphere, representing a ship's compass. It was originally regarded as part of Argo.

quadrature
Position of the moon or an outer planet at which a line between it and Earth makes a right angle with a line joining Earth to the sun.

quasar (from quasi-stellar object, or QSO)
One of the most distant extragalactic objects known, first discovered in 1963. Quasars appear starlike, but each emits more energy than one hundred giant galaxies. They are thought to be at the center of galaxies, their brilliance emanating from the stars and gas falling toward an immense black hole at their nucleus. Quasar light shows a large red shift, indicating that they are very distant. Some

quasars emit radio waves, which is how they were first identified, but most are radio quiet. The farthest are more than 10 billion light-years away.

radar astronomy

Bouncing of radio waves off objects in the solar system, with reception and analysis of the "echoes." Radar contact was first made with the moon in 1945 and with Venus in 1961. The travel time for radio reflections allows the distances of objects to be determined accurately. Analysis of the reflected beam reveals the rotation period and allows the object's surface to be mapped. The rotation periods of Venus and Mercury were first determined by radar. Radar maps of Venus were obtained first by Earth-based radar and subsequently by orbiting space probes.

radial velocity

The velocity of an object, such as a star or a galaxy, along the line of sight, moving toward or away from an observer. The amount of Doppler shift (apparent change in wavelength) of the light reveals the object's velocity. If the object is approaching, the Doppler effect causes a blue shift in its light—that is, the wavelengths of light coming from the object appear to be shorter, tending toward the blue end of the spectrum. If the object is receding, there is a red shift, meaning that the wavelengths appear to be longer, toward the red end of the spectrum.

radio astronomy

Study of radio waves emitted naturally by objects in space, by means of a radio telescope. Radio emission comes from hot gases (thermal radiation), electrons spiraling in magnetic fields (synchrotron radiation), and specific wavelengths (lines) emitted by atoms and molecules in space, such as the 21-cm/8.3-in line emitted by hydrogen gas.

Radio astronomy began in 1932 when U.S. radio astronomer Karl Jansky detected radio waves from the center of our galaxy, but the subject did not develop until after World War II. Radio astronomy has greatly improved our understanding of the evolution of stars, the structure of galaxies, and the origin of the universe. Astronomers have mapped the spiral structure of the Milky Way from the radio waves given out by interstellar gas, and they have detected many individual radio sources within our galaxy and beyond.

Among radio sources in our galaxy are the remains of supernova explosions, such as the Crab Nebula and pulsars. Short-wavelength radio waves have been detected from complex molecules in dense clouds of gas where stars are forming. Searches have been undertaken for signals from other civilizations in the galaxy, so far without success.

Strong sources of radio waves beyond our galaxy include radio galaxies and quasars. Their existence far off in the universe demonstrates how the universe has evolved with time. Radio astronomers have also detected weak cosmic background radiation thought to be from the Big Bang explosion that marked the birth of the universe.

radio galaxy

Galaxy that is a strong source of electromagnetic waves of radio wavelengths. All galaxies, including our own, emit some radio waves, but radio galaxies are up to a million times more powerful. In many cases, the strongest radio emission comes not from the visible galaxy but from two clouds, invisible through an optical telescope, that can extend for millions of light-years either side of the galaxy. This double structure at radio wavelengths is also shown by some quasars, suggesting a close relationship between the two types of object. In both cases, the source of energy is thought to be a massive black hole at the center. Some radio galaxies are thought to result from two galaxies in collision or recently merged.

radio telescope

Instrument for detecting radio waves from the universe in radio astronomy. Radio telescopes usually consist of a metal bowl that collects and focuses radio waves the way a concave mirror collects and focuses light waves. Radio telescopes are much larger than optical telescopes, because the wavelengths they are detecting are much longer than the wavelength of light. The largest single dish is 305 m/ 1,000 ft across, at Arecibo, Puerto Rico.

A large dish, such as that at Jodrell Bank, Cheshire, England, can see the radio sky less clearly than a small optical telescope sees the visible sky. Interferometry is a technique in which the output from two dishes is combined to give better resolution of detail than with a single dish. Very long baseline interferometry (VBLI) uses radio telescopes spread across the world to resolve minute details of radio sources.

In aperture synthesis, several dishes are linked together to simulate the performance of a very large single dish. This technique was pioneered by British radio astronomer Martin Ryle at the Mullard Radio Astronomy Observatory, Cambridge, England, site of a radio telescope consisting of eight dishes in a line 5 km/ 3 mi long. The Very Large Array in New Mexico, consists of twenty-seven dishes arranged in a Y-shape, which simulates the performance of a single dish 27 km/ 17 mi in diameter. Other radio telescopes are shaped like long troughs, and some consist of simple rod-shaped aerials.

red dwarf

Any star that is cool, faint, and small (with a mass and a diameter about 0.1 the mass and the diameter of the sun). Red dwarfs burn slowly and have estimated lifetimes of 100 billion years. They may be the most abundant type of star but are difficult to see because they are so faint. Two of the closest stars to the sun, Proxima Centauri and Barnard's Star, are red dwarfs.

red giant

Any large bright star with a cool surface. It is thought to represent a late stage in the evolution of a star like the sun, as it runs out of hydrogen fuel at its center and begins to burn heavier elements, such as helium, carbon, and silicon. Because of more complex nuclear reactions that then occur in the red giant's interior, it eventually becomes gravitationally unstable and begins to collapse and heat up. The result is either explosion of the star as a supernova, leaving behind a neutron star, or loss of mass by more gradual means to produce a white dwarf.

Red giants have diameters between ten and one hundred times that of the sun. They are very bright because they are so large, although their surface temperature is lower than that of the sun, about 2,000–3,000 K (1,700–2,700°C/ 3,000–5,000°F).

red shift

The lengthening of the wavelengths of light from an object as a result of the object's motion away from us. It is an example of the Doppler effect. The red shift in light from galaxies is evidence that the universe is expanding.

Lengthening of wavelengths causes the light to move or shift toward the red end of the spectrum; hence, the name. The amount of red shift can be measured by the displacement of lines in an object's spectrum. By measuring the amount of red shift in light from stars and galaxies, astronomers can tell how quickly these objects are moving away from us. A strong gravitational field can also produce a red shift in light; this is termed gravitational red shift.

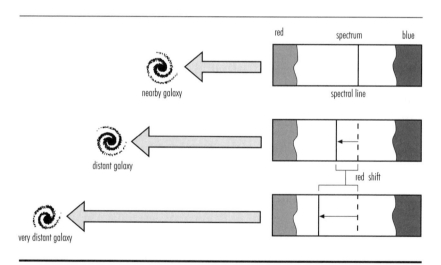

Redstone rocket

Short-range U.S. military missile, modified for use as a space launcher. *Redstone* rockets launched the first two flights of the Mercury Project. A modified Redstone, *Juno 1,* launched the first U.S. satellite, *Explorer 1,* in 1958.

reflecting telescope

A telescope in which light is collected and brought to a focus by a concave mirror. Cassegrain telescopes and Newtonian telescopes are examples.

refractor

In astronomy, a telescope in which light is collected and brought to a focus by a convex lens (the object lens, or objective).

Regulus, or Alpha Leonis

Brightest star in the constellation Leo and the twenty-first-brightest star in the sky. First-magnitude Regulus has a true luminosity one hundred times that of the sun and is 85 light-years from Earth. Regulus was one of the four royal stars of ancient Persia marking the approximate positions of the sun at the equinoxes and the solstices. The other three were Aldebaran, Antares, and Fomalhaut.

Reticulum

Small constellation of the Southern Hemisphere, represented as a net or a reticule (the grid of lines on the eyepiece of a telescope). It used to be called Rhombus or Reticulum Rhomboidalis.

retrograde

The orbit or rotation of a planet or a satellite when the sense of rotation is opposite the general sense of rotation of the solar system. On the celestial sphere, it refers to motion from east to west against the background of stars.

Rigel, or Beta Orionis

Brightest star in the constellation Orion. It is a blue-white supergiant with an estimated diameter fifty times that of the sun. It is 910 light-years from Earth. It is the seventh-brightest star in the sky; intrinsically the brightest of the first-magnitude stars, its luminosity is about 100,000 times that of the sun, Its name is derived from the Arabic for "foot."

right ascension

The coordinate on the celestial sphere that corresponds to longitude on the surface of Earth. It is measured in hours, minutes, and seconds eastward from the point at which the sun's path (the ecliptic) once a year intersects the celestial equator; this point is called the vernal equinox.

Rigil Kent

See Alpha Centauri, or Rigil Kent.

Roche limit

The distance from a planet within which a large moon would be torn apart by the planet's gravitational force, creating a set of rings. The Roche limit lies at approximately 2.5 times the planet's radius (the distance from its center to its surface).

rocket

Projectile driven by the reaction of gases produced by a fast-burning fuel. Unlike jet engines, which are also reaction engines, modern rockets carry their own

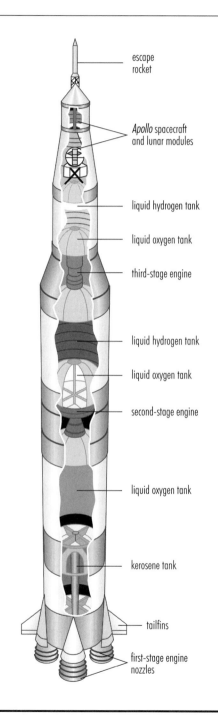

escape
rocket

Apollo spacecraft
and lunar modules

liquid hydrogen tank

liquid oxygen tank

third-stage engine

liquid hydrogen tank

liquid oxygen tank

second-stage engine

liquid oxygen tank

kerosene tank

tailfins

first-stage engine
nozzles

oxygen supply to burn their fuel and do not require any surrounding atmosphere. For warfare, rocket heads carry an explosive device.

Rockets have been valued as fireworks since the Middle Ages, but their intensive development as a means of propulsion to high altitudes, carrying payloads, started only in the interwar years with the state-supported work in Germany (primarily by German-born U.S. rocket engineer Wernher von Braun) and that of U.S. inventor Robert Hutchings Goddard in the United States. Because they are the only form of propulsion available that can function in a vacuum, rockets, including multistage ones that consist of a number of rockets joined together, are essential to exploration in outer space.

Two main kinds of rocket are used: One burns liquid propellants; the other, solid propellants. The fireworks rocket uses gunpowder as a solid propellant. The space shuttle's solid rocket boosters use a mixture of powdered aluminum in a synthetic rubber binder. Most rockets, however, have liquid propellants, which are more powerful and easier to control. Liquid hydrogen and kerosene are common fuels, while liquid oxygen is the most common oxygen provider, or oxidizer. One of the biggest rockets ever built, the *Saturn V* moon rocket, was a three-stage design, standing 111 m/365 ft high. It weighed more than 2,700 metric tons/3,000 tons on the launch pad, developed a takeoff thrust of 3.4 million kg/7.5 million lb, and could place almost 140 metric tons/150 tons into low Earth orbit. In the early 1990s, the most powerful rocket system was the Soviet *Energiya,* capable of placing 190 metric tons/210 tons into low Earth orbit. The U.S. space shuttle can only carry up to 29 metric tons/32 tons of equipment.

ROSAT
Joint U.S.-German-British satellite launched in 1990 to study cosmic sources of X rays and extremely short ultraviolet wavelengths; the satellite was named for German physicist Wilhelm Röntgen, the discoverer of X rays.

Rosetta
In astronomy, a project of the European Space Agency, due for launch in 2003, to send a spacecraft to Comet Wirtanen. *Rosetta* is expected to go into orbit around the comet in 2011 and land two probes on the nucleus a year later. The spacecraft will stay with the comet as it makes its closest approach to the sun in October 2013.

Royal Greenwich Observatory
The national astronomical observatory of Britain, founded in 1675 at Greenwich, England, to provide navigational information for sailors. From 1990 until its closure in 1998 it was based in Cambridge, England.

Sagitta
Small constellation of the Northern Hemisphere, represented as an arrow. It lies in the Milky Way.

Sagittarius

Bright zodiac constellation in the Southern Hemisphere, represented as an archer aiming a bow and arrow at neighboring Scorpius. The sun passes through Sagittarius from mid-December to mid-January, including the winter solstice, when it is farthest south of the equator. The constellation contains many nebulae, globular clusters, and open star clusters. Kaus Australis and Nunki are its brightest stars. The center of our galaxy, the Milky Way, is marked by the radio source Sagittarius A. In astrology, the dates for Sagittarius are about November 22–December 21.

Salyut

Russian for "salute"; a series of seven space stations launched by the USSR (1971–1986). *Salyut* was cylindrical in shape, 15 m/50 ft long, and weighed 19 metric tons/21 tons. It housed two or three cosmonauts at a time for missions lasting up to eight months.

Salyut 1 was launched April 19, 1971. It was occupied for twenty-three days in June 1971 by a crew of three, who died during their return to Earth when their *Soyuz* ferry craft depressurized. In 1973, *Salyut 2* broke up in orbit before occupation. The first fully successful *Salyut* mission was a fourteen-day visit to *Salyut 3* in July 1974. In 1984–1985, a team of three cosmonauts endured a record 237-day flight in *Salyut 7*. In 1986, the *Salyut* series was superseded by *Mir,* an improved design capable of being enlarged by additional modules sent up from Earth.

Crews observed Earth and the sky and carried out the processing of materials in weightlessness. The last in the series, *Salyut 7,* crashed to Earth in February 1991, scattering debris over Argentina.

satellite

Any small body, either natural or artificial, that orbits a larger one. Natural satellites that orbit planets are called moons. The first artificial satellite, *Sputnik 1,* was launched into orbit around Earth in 1957 by the USSR. Artificial satellites are used for scientific purposes, communications, weather forecasting, and military applications. The brightest artificial satellites can be seen by the naked eye.

At any time, there are several thousand artificial satellites orbiting Earth, including active satellites, satellites that have ended their working lives, and discarded sections of rockets. Artificial satellites eventually reenter Earth's atmosphere. Usually, they burn up by friction, but sometimes debris falls to Earth's surface, as with the *Skylab* and *Salyut 7* space stations. In 1997, there were three hundred active artificial satellites in orbit around Earth, the majority used in communications.

Saturn

The second-largest planet in the solar system, sixth from the sun, and encircled by bright and easily visible equatorial rings. Viewed through a telescope, it is ocher. Its polar diameter is 12,000 km/7,450 mi smaller than its equatorial diameter, a result of its fast rotation and low density, the lowest of any planet. Its mass is ninety-five times that of Earth, and its magnetic field a thousand times stronger.

mean distance from the sun:
1.427 billion km/0.886 billion mi

equatorial diameter:
120,000 km/75,000 mi

rotational period:
0.43 Earth day at equator, 0.44 Earth day at higher latitudes

year:
29.46 Earth years

atmosphere:
Visible surface consists of swirling clouds, probably made of ammonia crystals at a temperature of –170°C/–274°F, although the markings in the clouds
are not as prominent as Jupiter's. The U.S. space probes *Voyager 1* and *2* found
winds reaching 1,800 kph/1,100 mph.

surface:
Saturn is believed to have a small core of rock and iron, encased in ice and
topped by a deep layer of liquid hydrogen.

satellites:
Eighteen known moons, more than for any other planet. The largest moon,
Titan, has a dense atmosphere. Other satellites include Epimetheus, Janus, Pan-
dor, and Prometheus.

rings:
The rings visible from Earth begin about 14,000 km/9,000 mi from the planet's
cloud tops and extend out to about 76,000 km/47,000 mi. Made of small chunks
of ice and rock (averaging 1 m/3.3 ft across), they are 275,000 km/170,000 mi
rim to rim but only 100 m/300 ft thick. The *Voyager* probes showed that the
rings actually consist of thousands of closely spaced ringlets, looking like the
grooves in a phonograph record.

    From Earth, Saturn's rings appear to be divided into three main sections.
Ring A, the outermost, is separated from ring B, the brightest, by the Cassini
division, named for its seventeenth-century discoverer, Italian astronomer
Giovanni Cassini, which is 3,000 km/2,000 mi wide; the inner, transparent
ring C is also called the Crepe Ring. Each ringlet of the rings is made of a
swarm of icy particles like snowballs, a few centimeters to a few meters in
diameter. Outside the A ring is the narrow and faint F ring, which the
Voyagers showed to be twisted or braided. The rings of Saturn could be
the remains of a shattered moon, or they may always have existed in their
present form.

Saturn rocket

Family of large U.S. rockets, developed by German-born U.S. rocket engineer Wernher von Braun for the Apollo Project. The two-stage *Saturn IB* was used for launching *Apollo* spacecraft into orbit around Earth. The three-stage *Saturn V* sent *Apollo* spacecraft to the moon and launched the *Skylab* space station. The takeoff thrust of a *Saturn V* was 3.4 million kg/7.5 million lb. After *Apollo* and *Skylab,* the *Saturn* rockets were retired in favor of the space shuttle.

Schmidt Telescope

Reflecting telescope used for taking wide-angle photographs of the sky. Invented in 1930 by Estonian astronomer Bernhard Schmidt, it has an added corrector lens to help focus the incoming light. Examples are the 1.2-m/48-in Schmidt Telescope on Mount Palomar, near Pasadena, California, and the UK Schmidt Telescope, of the same size, at the Anglo-Australian Observatory at Siding Spring Mountain, New South Wales, Australia.

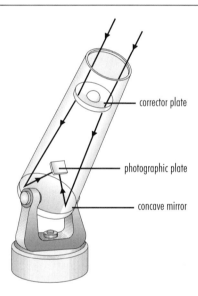

corrector plate

photographic plate

concave mirror

Schwarzschild radius

In astrophysics, the radius of the event horizon surrounding a black hole within which light cannot escape its gravitational pull. For a black hole of mass m, the Schwarzschild radius Rs is given by Rs = $2gm/c$, where g is the gravitational constant and c is the speed of light. The Schwarzschild radius for a black hole of solar mass is about 3 km/1.9 mi. It is named for Karl Schwarzschild, the German astronomer who deduced the possibility of black holes from Einstein's general theory of relativity in 1916.

Scorpio
Astrological term for Scorpius.

Scorpius
Bright zodiacal constellation in the Southern Hemisphere between Libra and Sagittarius, represented as a scorpion. The sun passes briefly through Scorpius in the last week of November. The heart of the scorpion is marked by the bright red supergiant star Antares. Scorpius contains rich Milky Way star fields, plus the strongest X-ray source in the sky, Scorpius X-1. The whole area is rich in clusters and nebulae. In astrology, the dates for Scorpius are about October 24–November 21.

Sculptor
Inconspicuous constellation of the Southern Hemisphere, containing the south pole of the galaxy. It was originally named by French astronomer Nicolas Lacaille as Apparatus Sculptoris, "the Sculptor's Tools."

Scutum
Small constellation of the Southern Hemisphere, represented by a shield. It lies in the Milky Way, which in this region consists of uniform clouds of faint stars.

Serpens
Constellation on the celestial equator, represented as a serpent coiled around the body of Ophiuchus. It is the only constellation divided into two halves: Serpens Caput, the head (on one side of Ophiuchus), and Serpens Cauda, the tail (on the other side). Its main feature is the Eagle Nebula.

SETI (search for extraterrestrial intelligence)
A program originally launched in 1992 by the U.S. National Aeronautics and Space Administration (NASA), using powerful radio telescopes to search the skies for extraterrestrial signals. NASA canceled the SETI Project in 1993, but other privately funded SETI projects continue.

Sextans
Faint constellation on the celestial equator, represented as a sextant.

Seyfert galaxy
Galaxy whose small, bright center is caused by hot gas moving at high speed around a massive central object, possibly a black hole. Almost all Seyferts are spiral galaxies. They seem to be closely related to quasars but are about one hundred times fainter. They are named for their discoverer, U.S. astronomer and astrophysicist Carl Seyfert.

shooting star
See meteor.

sidereal period
The orbital period of a planet around the sun, or a moon around a planet, with reference to a background star. The sidereal period of a planet is, in effect, a "year" for that planet. A synodic period is a full circle as seen from Earth.

sidereal time
Time measured by the rotation of Earth with respect to the stars. A sidereal day is the time taken by Earth to turn once with respect to the stars—namely, 23.93 hours. It is divided into sidereal hours, minutes, and seconds, each of which is proportionally shorter than the corresponding SI (metric) unit.

Siding Spring Mountain
Peak 400 km/250 mi northwest of Sydney, Australia, site of the Anglo-Australian Observatory holding the 1.2 m/48-in UK Schmidt Telescope, opened in 1973, and the 3.9-m/154-in Anglo-Australian Telescope, opened in 1975, which was the first big telescope to be fully computer controlled. It is one of the most powerful telescopes in the Southern Hemisphere.

Sigma Octantis
The star closest to the south celestial pole, in effect the southern equivalent of Polaris, although far less conspicuous. Situated just less than 1° from the south celestial pole in the constellation Octans, Sigma Octantis is 120 light-years away from Earth.

singularity
In astrophysics, the point in space-time at which the known laws of physics break down. Singularity is predicted to exist at the center of a black hole, where infinite gravitational forces compress the infalling mass of a collapsing star to infinite density. It is also thought, according to the Big Bang model of the origin of the universe, to be the point from which the expansion of the universe began.

Sirius, or Dog Star, or Alpha Canis Majoris
Brightest star in the night sky, 8.6 light-years from Earth in the constellation Canis Major. Sirius is a double star: Sirius A is a white star with a mass 2.3 times that of the sun, a diameter 1.8 times that of the sun, and a luminosity of twenty-three suns. It is orbited every fifty years by a white dwarf, Sirius B, also known as the Pup. Sirius B ia an eighth-magnitude star sometimes known as "the Dark Companion," as it was first detected in the nineteenth century by German astronomer Friedrich Bessel from its gravitational effect on the proper motion of Sirius A.

   "Dog Star" is the alternative name for Sirius, dating back to ancient Egypt. The unpleasantness of the hot summer season known as the "dog days" was attributed to the influence of Sirius being in conjunction with the sun.

Skylab

U.S. space station, launched May 14, 1973, made from the adapted upper stage of a *Saturn V* rocket. At 75 metric tons/82.5 tons, it was the heaviest object ever put into space and was 25.6 m/84 ft long. *Skylab* contained a workshop for carrying out experiments in weightlessness, an observatory for monitoring the sun, and cameras for photographing Earth's surface.

Damaged during launch, it had to be repaired by the first crew of astronauts. Three crews, each of three astronauts, occupied *Skylab* for periods of up to eighty-four days, at that time a record duration for human spaceflight. *Skylab* finally fell to Earth July 11, 1979, dropping debris over western Australia.

SNR

See supernova remnant (SNR).

SOHO *(Solar and Heliospheric Observatory)*

Space probe launched in 1995 by the European Space Agency to study the solar wind of atomic particles streaming toward Earth from the sun. It also observes the sun in ultraviolet and visible light and measures slight oscillations on the sun's surface that can reveal details of the structure of the sun's interior. It is positioned 1.5 million km/938,000 mi from Earth toward the sun. SOHO is operated jointly with the U.S. National Aeronautics and Space Administration (NASA) and costs $1.2 billion.

SOHO carries equipment for eleven separate experiments, including the study of the sun's corona, measurement of its magnetic field, and of solar winds. The Coronal Diagnostic Spectrometer (CDS) detects radiation at extreme ultraviolet wavelengths and allows the study of the sun's atmosphere. The Michelson Doppler Imager (MDI) measures Doppler shifts in light wavelengths and can detect winds caused by convention beneath the sun's surface. The Extreme-Ultraviolet Imaging Telescope (EIT) investigates the mechanisms that heat the sun's corona. The Large-Angle Spectroscopic Coronagraph (LASCO) images the corona by detecting sunlight scattered by the coronal gases.

Solar and Heliospheric Observatory

See SOHO *(Solar and Heliospheric Observatory)*.

solar cycle

The variation of activity on the sun over an eleven-year period indicated primarily by the number of sunspots visible on its surface. The next period of maximum activity is expected around 2001.

solar flare

Brilliant eruption on the sun above a sunspot, thought to be caused by release of magnetic energy. Flares reach maximum brightness within a few minutes, then fade away over about an hour. They eject a burst of atomic particles into space at up to 1,000 kps/600 mps. When these particles reach Earth, they can cause radio blackouts, disruptions of Earth's magnetic field, and aurorae.

Solar Maximum Mission

Satellite launched by the U.S. National Aeronautics and Space Administration (NASA) in 1980 to study solar activity, which discovered that the sun's luminosity increases slightly when sunspots are most numerous. It was repaired in orbit by astronauts from the space shuttle in 1984 and burned up in Earth's atmosphere in 1989.

solar spicules

Short-lived jets of hot gas in the upper chromosphere of the sun. Spiky in appearance, they move at high velocities along lines of magnetic force, to which they owe their shape, and last for a few minutes each. Spicules appear to disperse material into the corona.

solar system

The sun (a star) and all of the bodies orbiting it: the nine planets (Mercury, Venus, Earth, Mars, Jupiter, Saturn, Uranus, Neptune, and Pluto), their moons, the asteroids, and the comets. The sun contains 99.86 percent of the mass of the solar system.

    The solar system gives every indication of being a strongly unified system having a common origin and development. It is isolated in space; all of the planets go round the sun in orbits that are nearly circular and coplanar and in the same direction as the sun itself rotates; moreover, this same pattern is continued in the regular system of satellites that accompany Jupiter, Saturn, and Uranus. The solar system is thought to have formed by condensation from a cloud of gas and dust in space about 4.6 billion years ago.

solar time

The time of day as determined by the position of the sun in the sky. Apparent solar time, the time given by a sundial, is not uniform because of the varying speed of Earth in its elliptical orbit. Mean solar time is a uniform time that coincides with apparent solar time at four instants through the year. The difference between them is known as the equation of time and is greatest in early November when the sun is more than sixteen minutes fast on mean solar time. Mean solar time on the Greenwich meridian (the zero degree line of longitude which passes through the Old Royal Observatory at Greenwich, London) is known as Greenwich Mean Time (GMT) and is the basis of civil timekeeping.

solar wind

Stream of atomic particles, mostly protons and electrons, from the sun's corona, flowing outward at speeds of between 300 kps/200 mps and 1,000 kps/600 mps. The fastest streams come from "holes" in the sun's corona that lie over areas where no surface activity occurs. The solar wind pushes the gas of comets' tails away from the sun, and "gusts" in the solar wind cause geomagnetic disturbances and aurorae on Earth.

solstice
Either of the days on which the sun is farthest north or south of the celestial equator each year. The summer solstice, when the sun is farthest north, occurs around June 21; the winter solstice, when it is farthest south, around December 22.

South African Astronomical Observatory
National observatory of South Africa at Sutherland, founded in 1973 after the merger of the Royal Observatory, Cape Town, and the Republic Observatory, Johannesburg, and operated by the Council for Scientific and Industrial Research of South Africa. Its main telescope is a 1.88-m/74-in reflector formerly at the Radcliffe Observatory, Pretoria.

Southern Cross
See Crux.

southern lights
See aurora.

Soyuz
Soviet (now Russian) series of spacecraft, capable of carrying up to three cosmonauts. *Soyuz* spacecraft consist of three parts: a rear section that contains the engines; the central crew compartment; and a forward compartment that gives additional room for working and living space. They are now used for ferrying crews up to space stations, though they were originally used for independent space flight. *Soyuz 1* crashed on its first flight in April 1967, killing the lone pilot, Vladimir Komarov. Three cosmonauts were killed in 1971 aboard *Soyuz 11,* while returning from a visit to the *Salyut 1* space station, when a faulty valve caused their cabin to lose pressure. In 1975, the joint U.S.-Soviet *Apollo-Soyuz* test project resulted in a successful docking of the two spacecraft in orbit.

space, or outer space
The void that exists beyond Earth's atmosphere. Above 120 km/75 mi, very little atmosphere remains, so objects can continue to move quickly without extra energy. The space between the planets is not entirely empty; it is filled with the tenuous gas of the solar wind, as well as dust specks.

space-adaptation syndrome
See space sickness, or space-adaptation syndrome.

space probe
Any instrumented object sent beyond Earth to collect data from other parts of the solar system and from deep space. The first probe was the Soviet *Lunik 1,*

which flew past the moon in 1959. The first successful planetary probe was the U.S. *Mariner 2,* which flew past Venus in 1962, using transfer orbit. The first space probe to leave the solar system was the U.S. *Pioneer 10* in 1983. Space probes include *Galileo, Giotto, Magellan, Mars Observer, Ulysses,* the moon probes, and the *Mariner, Pioneer, Viking,* and *Voyager* series.

space shuttle

Reusable crewed spacecraft. The first was launched on April 12, 1981, by the United States. It was developed by the National Aeronautics and Space Administration (NASA) to reduce the cost of using space for commercial, scientific, and military purposes. After leaving its payload in space, the space-shuttle orbiter can be flown back to Earth to land on a runway and is then available for reuse.

Four orbiters were built initially: *Columbia, Challenger, Discovery,* and *Atlantis. Challenger* was destroyed in a midair explosion just more than a minute after its tenth launch, on January 28, 1986, killing all seven crew members, the result of a failure in one of the solid rocket boosters. Flights resumed with redesigned boosters in September 1988. A replacement orbiter, *Endeavor,* was built, which had its maiden flight in May 1992. At the end of the 1980s, an average of $375 million had been spent on each space-shuttle mission.

The USSR produced a shuttle of similar size and appearance to the U.S. one. The first Soviet shuttle, *Buran,* was launched without a crew by the *Energiya* rocket on November 15, 1988. In Japan, development of a crewless shuttle began in 1986.

The U.S. space-shuttle orbiter, the part that goes into space, is 37.2 m/122 ft long and weighs 68 metric tons/75 tons. Two to eight crew members occupy the orbiter's nose section, and missions last up to thirty days. In its cargo bay, the orbiter can carry up to 29 metric tons/32 tons of satellites, scientific equipment, *Spacelab,* or military payloads. At launch, the shuttle's three main engines are fed with liquid fuel from a cylindrical tank attached to the orbiter; this tank is discarded shortly before the shuttle reaches orbit. Two additional solid-fuel boosters provide the main thrust for launch but are jettisoned after two minutes.

space sickness, or space-adaptation syndrome

Feeling of nausea, sometimes accompanied by vomiting, experienced by about 40 percent of all astronauts during their first few days in space. It is akin to travel sickness and is thought to be caused by confusion of the body's balancing mechanism, located in the inner ear, by weightlessness. The sensation passes after a few days as the body adapts.

space station

Any large structure designed for human occupation in space for extended periods of time. Space stations are used for carrying out astronomical observations and surveys of Earth, as well as for biological studies and the processing of materials in weightlessness. The first space station was the Soviet *Salyut 1,* launched in 1971. In 1973, the United States launched *Skylab,* which was visited by three different astronaut crews in the 1970s before falling to Earth in mid-1979.

The U.S. National Aeronautics and Space Administration (NASA) plans to build a larger space station, to be called *Alpha,* in cooperation with other countries, including the European Space Agency; Russia and Japan are also building modules. The Russian-built control module *Zarya* was the first component launched (November 20, 1998) for the Internnational Space Station. It is one of the U.S. components. Once the *Zarya* reached orbit, the space shuttle *Endeavour* (launched December 3, 1998) made a rendezvous and attached a U.S.-built connecting module called *Unity.* The third component, a Russian provided crew living quarters and an early station core known as the service module will be the next component launched (scheduled for May 1999).

space suit
Protective suit worn by astronauts and cosmonauts in space. It provides an insulated, air-conditioned cocoon in which people can live and work for hours at a time while outside the spacecraft. Inside the suit is a cooling garment that keeps the body at a comfortable temperature even during vigorous work. The suit provides air to breathe and removes exhaled carbon dioxide and moisture. The suit's outer layers insulate the occupant from the extremes of hot and cold in space (–150°C/–240°F in the shade to +180°C/+350°F in sunlight) and from the impact of small meteorites. Some space suits have a jet-propelled backpack, which the wearer can use to move about.

Spacelab
Small space station built by the European Space Agency, carried in the cargo bay of the U.S. space shuttle, in which it remains throughout each flight, returning to Earth with the shuttle. *Spacelab* consists of a pressurized module in which astronauts can work and a series of pallets, open to the vacuum of space, on which equipment is mounted.

Spacelab is used for astronomy, Earth observation, and experiments utilizing the conditions of weightlessness and vacuum in orbit. The pressurized module can be flown with or without pallets, or the pallets can be flown on their own, in which case the astronauts remain in the shuttle's own crew compartment. All sections of *Spacelab* can be reused many times. The first *Spacelab* mission, consisting of a pressurized module and pallets, lasted ten days in November–December 1983.

speckle interferometry
Technique whereby large telescopes can achieve high resolution of astronomical objects despite the adverse effects of the atmosphere, through which light from the object under study must pass. It involves taking large numbers of images, each under high magnification and with short exposure times. The pictures are then combined to form the final picture.

spectral classification
The classification of stars according to their surface temperature and luminosity, as determined from their spectra. Stars are assigned a spectral type (or class)

denoted by the letters O, B, A, F, G, K, and M, in which O stars (about 40,000 K/ 39,700°C/71,500°F) are the hottest and M stars (about 3,000 K/2,700°C/5,000°F) are the coolest.

Each letter may be further divided into ten subtypes, B0, B1, B2, and so on. Stars are also assigned a luminosity class denoted by a Roman numeral attached to the spectral type: I (supergiants), II (bright giants), III (giants), IV (subgiants), V (main sequence), VI (subdwarfs), or VII (white dwarfs). The sun is classified as type G2V.

spectroscopic binary

A binary star in which two stars are so close together that they cannot be seen separately, but their separate light spectra can be distinguished by a spectroscope.

The first spectroscopic binary to be discovered, in 1889 by U.S. astronomer Edward Pickering, was the brighter component of Mizar. Many hundreds are now known, with orbital periods ranging from eighty-two minutes to fifteen years. The most common period is a few days.

As the two stars revolve around their mutual center of mass, they alternately approach and recede from the observer, resulting in a periodic Doppler shift in the lines of their spectra. If the orbital motion happens to lie at right angles to the line of sight, there is no approach or recession, so that such a system cannot be detected as a spectroscopic binary. In about one case in six, the component stars are sufficiently similar in brightness for the spectra of both to appear; then, as one star approaches, the other recedes, so that the spectral lines appear sometimes double, sometimes single, giving a double-line spectroscopic binary.

The line-of-sight velocity of the brighter star, or of each star in a double-line spectroscopic binary, is measured from the Doppler shift at various stages of the orbital period. Analysis of these velocity curves gives a lower limit of the combined mass of a single-line binary or of the mass of each component of a double-line binary.

Spica, or Alpha Virginis

Brightest star in the constellation Virgo and the sixteenth-brightest star in the sky. First-magnitude Spica has a true luminosity of more than fifteen hundred times that of the sun and is 260 light-years from Earth. It is also a spectroscopic binary star, the components of which orbit each other every four days. Spica is Latin for "ear of corn."

spicules, solar

See solar spicules.

spiral galaxy

One of the main classes of galaxy in the Hubble classification comprising up to 30 percent of known galaxies. Spiral galaxies are characterized by a central bulge surrounded by a flattened disk containing (normally) two spiral arms composed of hot young stars and clouds of dust and gas. In about half of spiral galaxies (barred spirals), the arms originate at the ends of a bar across the central bulge. The bar is not a rigid object but consists of stars in motion about the center of the galaxy.

Sputnik

Series of ten Soviet Earth-orbiting satellites. *Sputnik 1* was the first artificial satellite, launched October 4, 1957. It weighed 84 kg/184 lb, was 58 cm/23 in in diameter, and carried only a simple radio transmitter that allowed scientists to track it as it orbited Earth. It burned up in the atmosphere ninety-two days later. *Sputniks* were superseded in the early 1960s by the *Cosmos* series.

Sputnik 2, launched November 3, 1957, weighed about 500 kg/1,100 lb, including the dog Laika, the first living creature in space. Unfortunately, there was no way to return the dog to Earth, and it died in space. Later *Sputniks* were test flights of the *Vostok* spacecraft.

star

Luminous globe of gas, mainly hydrogen and helium, that produces its own heat and light by nuclear reactions. Although stars shine for a very long time—many billions of years—they are not eternal and have been found to change in appearance at different stages in their lives.

The smallest mass possible for a star is about 0.8 that of the sun (eighty times that of Jupiter); otherwise, nuclear reactions do not occur. Objects with less than this critical mass shine only dimly and are termed brown dwarfs.

origin

Stars are born when nebulae (giant clouds of dust and gas) contract under the influence of gravity. These clouds consist mainly of hydrogen and helium, with traces of other elements and dust grains. A huge volume of interstellar matter gradually separates from the cloud, and the temperature and pressure in its core rises as the star grows smaller and denser. As the star is forming, it is surrounded by evaporating gaseous globules (EGGs), the oldest of which was photographed in the Eta Carina Nebula in 1996 by the Hubble Space Telescope.

At first the temperature of the star scarcely rises, as dust grains radiate away much of the heat, but, as it grows denser, less of the heat generated can escape, and it gradually warms up. At about 10 million°C/18 million°F, the temperature is hot enough for a nuclear reaction to begin, and hydrogen nuclei fuse to form helium nuclei; vast amounts of energy are released, contraction stops, and the star begins to shine.

main-sequence stars

Stars at this stage are called main-sequence stars. Enough energy is produced in the nuclear reaction to replace that being lost at the surface, so the star has no need to contract further until its nuclear energy sources are exhausted. Until this happens, the star remains practically unaltered. Where the star is on the main sequence, how long it takes to contract before it gets there, and how long it remains there are all determined by the mass of the star; the larger the mass, the shorter the period and the brighter the star. A star with the mass of the sun takes a few million years to reach the main sequence, and it then remains on it for about ten billion years, a little more than twice the present age of the sun. The sun is, thus, expected to remain at this stage for another five billion years.

Surface temperatures of main-sequence stars range from 2,000°C/3,600°F to above 30,000°C/54,000°F. The corresponding colors range from red to blue-white. The nuclear reactions take place near the center, so the star gradually acquires an inert helium core, surrounded by a thin shell of burning hydrogen. When all of the hydrogen at the core of a main-sequence star has been converted into helium, the star swells to become a red giant, about one hundred times its previous size and with a cooler, redder surface.

white dwarfs
What happens next depends on the mass of the star. If this is less than 1.2 that of the sun, the star's outer layers drift off into space to form a planetary nebula, and its core collapses in on itself to form a small and very dense body called a white dwarf. Eventually, the white dwarf fades away, leaving a nonluminous dark body.

supernovae
If the mass is greater than 1.2 that of the sun, the star does not end as a white dwarf but passes through its life cycle quickly, becoming a red supergiant. As the star's core grows hotter, further nuclear transformations take place, resulting in the helium being converted, first, into carbon and oxygen, then into heavier elements, and, finally, into iron. The star eventually explodes into a brilliant supernova. Part of the core remaining after the explosion may collapse to form a small superdense star, consisting almost entirely of neutrons and, therefore, called a neutron star. Neutron stars, also called pulsars, spin very quickly, giving off pulses of radio waves.

black holes
If the collapsing core of the supernova has a mass more than three times that of the sun, it does not form a neutron star; instead, it forms a black hole, a region so dense that its gravity draws in not only all nearby matter, but also all radiation, including its own light, as the velocity of escape from its surface exceeds that of light. As French astronomer and mathematician Pierre Laplace pointed out in 1798, such a mass would not be visible from the outside. Black holes have been suggested as a possible explanation of various high-energy processes that are not yet understood, in quasars or X-ray sources such as Cygnus X-1.

star cluster
Group of related stars, usually held together by gravity. Members of a star cluster are thought to form together from one large cloud of gas in space. Open clusters,

such as the Pleiades, contain from a dozen to many hundreds of young stars, loosely scattered over several light-years. Globular clusters are larger and much more densely packed, containing perhaps from 10,000-100,000 old stars.

The more conspicuous clusters were originally cataloged with the nebulae and are usually known by their Messier or NGC numbers. A few clusters, like the Pleiades, the Hyades, and Praesepe, are also known by their traditional names. In the dense clusters, gravitation may hold the cluster stars together almost indefinitely, while, in the less dense clusters, the stars quickly dissipate.

globular clusters

About 120 globular clusters are known, all similar in appearance. They are nearly spherically symmetrical and contain an immense number of stars, which are concentrated toward the center, where they are so close together that it is impossible to see the individual stars even in the biggest telescopes. The brightest stars are red, and there are usually variable stars from which the distances of the clusters can be estimated. These distances range from 10,000 to 200,000 light-years, and it is possible that some of the more distant ones are not members of our galaxy. The diameters of the clusters range from 60 to 300 light-years, with 10,000–100,000 stars in a cluster, and an absolute magnitude in the 5–9 range. They are very old objects, possibly 10,000 million years old.

open clusters

The open clusters, of which the Pleiades, the Hyades, and Praesepe are typical, are much smaller than the globulars; they contain fewer stars, which are more widely scattered, and show little concentration toward the center. They are relatively young objects and are found in the disk of the galaxy, along the Milky Way. About a thousand of them have been identified, but there must be many more hidden by dust clouds or indistinguishable against the stellar background. The number of stars in an open cluster varies from a dozen to more than a thousand, but, on average, they contain a few hundred stars scattered over an area less than 30 light-years in diameter. The brightest stars are blue or red, depending on the cluster's age. Some of the youngest contain bright blue supergiant stars, which make the cluster's absolute magnitude as bright as 10, much brighter than the brightest globular, but the absolute magnitude of some of the open clusters can also be as faint as 0.

associations

Associations can be regarded as very extended open clusters with few members. They must have been recently formed or they would already have dispersed. About one hundred associations have been recognized in the spiral arms of the galaxy, but there must be many more hidden from view. A typical association contains up to fifty stars scattered irregularly over an area from 100 to 600 light-years in diameter. One of the best-known associations is that centered on a group of stars in the middle of the Orion Nebula.

As all members of a star cluster are about the same distance away, the observed relation between their colors and apparent magnitudes is essentially that between their colors and absolute magnitudes, which can be compared directly with the Hertzsprung-Russell diagram to measure the cluster's distance. Differences in the shape of this relationship from cluster to cluster provide evidence for the way stars evolve, as well as a method of determining a cluster's age.

starburst galaxy

A spiral galaxy that appears unusually bright in the infrared part of the spectrum due to a recent burst of star formation, possibly triggered by the gravitational influence of a nearby companion galaxy.

Stardust

U.S. project to obtain a sample of dust and gas from the head of a comet. Due for launch in February 1999, the *Stardust* space probe was scheduled to fly through the head of Comet Wild 2 in January 2004, passing within 100 km/62 mi of the 4 km/2.5 mi nucleus. It was to return to Earth with its samples in January 2006.

steady-state theory

In astronomy, a rival theory to that of the Big Bang, which posits that the universe has no origin but is expanding because new matter is being created continuously throughout the universe. The theory was proposed in 1948 by Austrian-born British cosmologist Hermann Bondi, Austrian-born U.S. astronomer Thomas Gold, and English astronomer, cosmologist, and writer Fred Hoyle, but it was dealt a severe blow in 1965 by the discovery of cosmic background radiation (radiation left over from the formation of the universe) and is now largely rejected.

stellar population

A classification of stars according to their chemical composition as determined by spectroscopy. Population I stars have a relatively high abundance of elements heavier than hydrogen and helium and are confined to the spiral arms and disk of the galaxy. They are believed to be young stars formed from material that has already been enriched with elements created by nuclear fusion in earlier generations of stars. Examples include open clusters and supergiants. Population II stars have a low abundance of heavy elements and are found throughout the galaxy but especially in the central bulge and the outer halo. They are among the oldest objects in the galaxy and include globular clusters. The sun is a Population II star.

sun

The star at the center of the solar system. Its diameter is 1.4 million km/865,000 mi; its temperature at the surface is about 5,800 K (5,530°C/9,986°F), and at the center 15 million K (about 15 million°C/27 million°F). It is composed of about 70 percent hydrogen and 30 percent helium, with other elements making up less than 1 percent. The sun's energy is generated by nuclear-fusion reactions that turn hydrogen into helium at its center. The gas core is far denser than mercury or lead on Earth. The sun is about 4.6 billion years old, with a predicted lifetime of ten billion years.

At the end of its life, it will expand to become a red giant the size of Mars's orbit, then shrink to become a white dwarf. The sun spins on its axis every twenty-five days near its equator but more slowly toward its poles. Its rotation can be followed by watching the passage of dark sunspots across its disk. Sometimes, bright eruptions called flares occur near sunspots.

Above the sun's photosphere (its visible surface which emits light and heat) lies a layer of thinner gas called the chromosphere, visible only by means of special instruments or at eclipses. Tongues of gas called prominences extend from the chromosphere into the corona, a halo of hot, tenuous gas surrounding the sun. Gas boiling from the corona streams outward through the solar system, forming the solar wind. Activity on the sun, including sunspots, flares, and prominences, waxes and wanes during the solar cycle, which peaks every eleven years or so and seems to be connected with the solar magnetic field.

A wall of heated hydrogen atoms at temperatures of 20,000–40,000 K (19,700–39,700°C/35,500–71,500°F) that forms in the path of the sun as it moves through space was discovered in 1995. The wall lies about 2,240 million km/1,555 million mi from the sun, and its existence had been predicted by theorists.

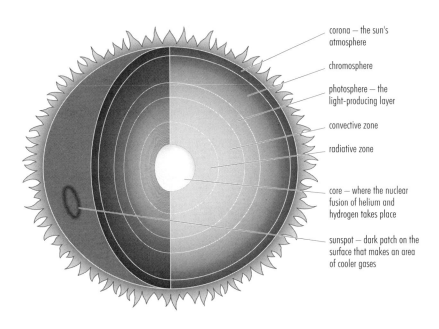

corona — the sun's atmosphere

chromosphere

photosphere — the light-producing layer

convective zone

radiative zone

core — where the nuclear fusion of helium and hydrogen takes place

sunspot — dark patch on the surface that makes an area of cooler gases

U.S. astronomers reported finding water on the sun in 1995. The water, in the form of superheated steam, was located in two sunspots where the temperature was only 3,300 K (3,000°C/5,400°F) (as opposed to 5,800 K [5,500°C/9,900°F] elsewhere on the surface).

sunspot

Dark patch on the surface of the sun, actually an area of cooler gas, thought to be caused by strong magnetic fields that block the outward flow of heat to the sun's surface. Sunspots consist of a dark central umbra, about 4,000 K (3,700°C/6,700°F), and a lighter surrounding penumbra, about 5,500 K (5,200°C/9,400°F). They last from several days to more than a month, ranging in size from 2,000 km/1,250 mi to groups stretching for more than 100,000 km/62,000 mi.

Sunspots are more common during active periods in the sun's magnetic cycle, when they are sometimes accompanied by nearby flares. The number of sunspots visible at a given time varies from none to more than a hundred, in a cycle averaging eleven years. There was a lull in sunspot activity, known as the Maunder minimum (1645–1715) that coincided with a cold spell in Europe.

Sunyaev-Zel'dovich effect

Slight dip in the cosmic background radiation when observed through the gas that surrounds clusters of galaxies. The size of the Sunyaev-Zel'dovich effect is a measure of the thickness of the gas through which the cosmic background radiation passes.

super cluster

A grouping of several clusters of galaxies to form a structure about 100–300 million light-years across. Our own galaxy and its neighbors lie on the edge of the "Local Super Cluster" of which the Virgo cluster is the dominant member.

supergiant

The largest and most luminous type of star known, with a diameter of up to a thousand times that of the sun and apparent magnitudes of between .4 and 1.3. Supergiants are likely to become supernovae.

superior planet

Planet that is farther away from the sun than Earth is. The superior planets are: Mars, Jupiter, Saturn, Uranus, Neptune, and Pluto.

supernova

The explosive death of a star, which temporarily attains a brightness of 100 million suns or more, so that it can shine as brilliantly as a small galaxy for a few days or weeks. Very approximately, it is thought that a supernova explodes in a large galaxy about once every one hundred years. Many supernovae—astronomers estimate some 50 percent—remain undetected because of obscuring by interstellar dust.

The name "supernova" was coined in 1934 by Swiss astronomer Fritz Zwicky and German-born U.S. astronomer Walter Baade. Zwicky was also responsible for the division into types I and II. Type I supernovae are thought to occur in binary-star systems, in which gas from one star falls on to a white dwarf, causing it to explode. Type II supernovae occur in stars ten or more times as massive as the sun, which suffer runaway internal nuclear reactions at the ends of their lives, leading to explosions. These are thought to leave behind neutron stars and black holes. Gas ejected by such an explosion causes an expanding radio source, such as the Crab Nebula. Supernovae are thought to be the main source of elements heavier than hydrogen and helium.

A supernova at maximum attains an absolute magnitude of between –14 and –20, more than ten thousand times brighter than an ordinary nova. When one occurs in a galaxy, its light quite often equals or exceeds the combined light of all of the other stars and nebulae in the system, while its spectrum indicates very violent internal motions. More than a dozen possible supernovae in our own galaxy that are still observable as radio sources have been identified among the "guest" stars recorded in Chinese and other annals. Of these, the best known are the one that appeared in Taurus in 1054 and gave rise to the Crab Nebula; the one observed in 1572 by Danish astronomer Tycho Brahe in Cassiopeia, which, for some days, outshone Venus and was visible in broad daylight; and the one described by both German mathematician and astronomer Johannes Kepler and the Italian astronomer and physicist Galileo that appeared in 1604 in Serpens. The first supernova was recorded (although not identified as such at the time) in A.D. 185 in China. The last supernova was seen in our galaxy in 1604, but many others have been seen since in other galaxies. In 1987, a supernova visible to the unaided eye occurred in the Large Magellanic Cloud, a small neighboring galaxy.

supernova remnant (SNR)
The glowing remains of a star that has been destroyed in a supernova explosion. The brightest and most famous example is the Crab Nebula.

synchronous orbit
See geostationary orbit, or synchronous orbit.

synchronous rotation
See captured rotation, or synchronous rotation.

synodic period
The time taken for a planet or a moon to return to the same position in its orbit as seen from Earth—that is, from one opposition to the next. It differs from the sidereal period because Earth is moving in orbit around the sun.

syzygy
The alignment of three celestial bodies, usually the sun, Earth, and the moon or the sun, Earth, and another planet. A syzygy involving the sun, Earth, and the

moon usually occurs during solar and lunar eclipses. The term also refers to the moon or another planet when it is in conjunction or opposition.

Tanegashima Space Center

Japanese rocket-launching site on a small island off South Kyushu. Tanegashima is run by the National Space Development Agency (NASDA), which is responsible for the practical applications of Japan's space program (research falls under a separate organization based at Kagoshima Space Center). NASDA, founded in 1969, has headquarters in Tokyo; a tracking and testing station, the Tsukuba Space Center, in east-central Honshu; and an Earth observation center near Tsukuba.

Tau Ceti

One of the nearest stars visible to the naked eye, 11.9 light-years from Earth in the constellation Cetus. It has a diameter slightly less than that of the sun and an actual luminosity of about 45 percent of the sun's. Its similarity to the sun is sufficient to suggest that Tau Ceti may possess a planetary system, although observations have yet to reveal evidence of this.

Taurus

Conspicuous zodiacal constellation in the Northern Hemisphere near Orion, represented as a bull. The sun passes through Taurus from mid-May to late June. In astrology, the dates for Taurus are about April 20–May 20.

The V-shaped Hyades open star cluster forms the bull's head, with Aldebaran as the red eye. The Pleiades open cluster is in the shoulder. Taurus also contains the Crab Nebula, the remnants of the supernova of A.D. 1054, which is a strong radio and X-ray source and the location of one of the first pulsars to be discovered.

In the conventional constellation figure, Taurus represents only the front part of a bull. The tips of the horns are marked by Beta and Zeta Tauri. Johann Bayer regarded Beta Tauri as belonging to Auriga and called it Gamma Aurigae. T. Tauri is one of a class of variable stars thought to represent an early stage of stellar formation.

tektite

From the Greek *tektos* ("molten"), a small, rounded glassy stone found in certain regions of Earth, such as Australasia. Tektites are probably the scattered drops of molten rock thrown out by the impact of a large meteorite.

telescope

Optical instrument that magnifies images of faint and distant objects; any device for collecting and focusing light and other forms of electromagnetic radiation. It is a major research tool in astronomy and is used to sight over land and sea; small telescopes can be attached to cameras and rifles. A telescope with a large aperture, or opening, can distinguish finer detail and fainter objects than one with a small aperture. The refracting telescope uses lenses, and the reflecting telescope uses mirrors. A third type, the catadioptric telescope, is a combination of lenses and mirrors.

refractor

In a refractor, light is collected by a lens called the object glass or objective, which focuses light down a tube, forming an image magnified by an eyepiece. The largest refracting telescope in the world, at Yerkes Observatory, Williams Bay, Wisconsin, has an aperture of 102 cm/40 in.

reflector

In a reflector, light is collected and focused by a concave mirror. Large mirrors are cheaper to make and easier to mount than large lenses, so all of the largest telescopes are reflectors. The largest reflector with a single mirror, 6 m/19.7 ft, is at Zelenchukskaya, Russia. Telescopes with larger apertures composed of numerous smaller segments have been built, such as the Keck Telescopes on Mauna Kea, Hawaii.. A multiple-mirror telescope was installed on Mount Hopkins, Arizona, in 1979. It originally consisted of six mirrors of 1.8 m/72 in aperture, which performed like a single 4.5-m/176-in mirror. The six mirrors were replaced in 1996 by a single 6.5-m/21.3-ft mirror. Schmidt telescopes are used for taking wide-field photographs of the sky. They have a main mirror plus a thin lens at the front of the tube to increase the field of view.

The liquid-mirror telescope is a reflecting telescope constructed with a rotating mercury mirror. In 1995, the U.S. National Aeronautics and Space Administration (NASA) completed a 3-m/9.8-ft liquid-mirror telescope at its Orbital Debris Observatory in New Mexico.

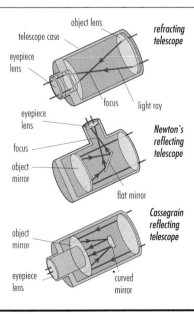

telescopes in space

Large telescopes can now be placed in orbit above the distorting effects of Earth's atmosphere. Telescopes in space have been used to study infrared, ultraviolet, and X-ray radiation that does not penetrate the atmosphere but carries much information about the births, lives, and deaths of stars and galaxies. The 2.4-m/94-in Hubble Space Telescope, launched in 1990 by the United States, can see the sky more clearly than can any telescope on Earth.

In 1996, an X-ray telescope was under development by British, U.S., and Australian astronomers, based on the structure of a lobster's eye, which has thousands of square tubes reflecting light onto the retina. The $6.4 million Lobster Eye Telescope will contain millions of tubes 10–20 micrometers across and is intended for use on a satellite. It is currently under technological development with construction scheduled to begin in 2001, and it is hoped to be launched by 2005.

Telescopium

Inconspicuous constellation of the Southern Hemisphere, represented as a telescope.

Telstar

U.S. communications satellite, launched July 10, 1962, which relayed the first live television transmissions between the United States and Europe. *Telstar* orbited Earth every 2.63 hours and so had to be tracked by ground stations, unlike the geostationary satellites of today.

terrestrial planet

Any of the four small, rocky inner planets of the solar system: Mercury, Venus, Earth, and Mars. The moon is sometimes also included, although it is a satellite of Earth and not strictly a planet.

tidal heating

A process in which one body is heated internally by tidal stresses set up by the gravitational pull of another body. Tidal heating is common among the moons of the giant planets and is the heat source for volcanic activity on Io, one of the moons of Jupiter.

Tidbinbilla

Space tracking station and nature reserve in Australia, just south of Canberra. It provides tracking facilities and command transmissions in support of U.S. National Aeronautics and Space Administration (NASA) crewed and uncrewed spacecraft, including the *Galileo* and *Cassini* probes.

Titan

Largest moon of the planet Saturn, with a diameter of 5,150 km/3,200 mi and a mean distance from Saturn of 1.2 million km/759,000 mi. It is the second-largest moon in the solar system (Ganymede, of Jupiter, is larger). Titan is the only moon in the solar system with a substantial atmosphere (mostly nitrogen); it is topped with smoggy orange clouds that obscure the surface, which may be covered with liquid-ethane lakes. Its surface atmospheric pressure is greater than Earth's. Radar signals suggest that Titan has dry land as well as oceans (among the planets, only Earth has both in the solar system).

Titan rocket

Family of U.S. space rockets, developed from the *Titan* intercontinental missile. Two-stage *Titan* rockets launched the *Gemini* crewed missions. More powerful *Titans*, with additional stages and strap-on boosters, were used to launch spy satellites and space probes, including the *Viking* and *Voyager* probes and the *Mars Observer*.

transfer orbit

Elliptical path followed by a spacecraft moving from one orbit to another, designed to save fuel, although at the expense of a longer journey time. Space probes travel to the planets on transfer orbits. A probe aimed at Venus has to be "slowed down" relative to Earth, so that it enters an elliptical transfer orbit with its perigee (point of closest approach to the sun) at the same distance as the orbit of Venus; toward Mars, the vehicle has to be "speeded up" relative to Earth, so that it reaches its apogee (farthest point from the sun) at the same distance as the orbit of Mars.

Geostationary transfer orbit is the highly elliptical path followed by satellites to be placed in geostationary orbit around Earth (an orbit coincident with Earth's rotation). A small rocket is fired at the transfer orbit's apogee to place the satellite in geostationary orbit.

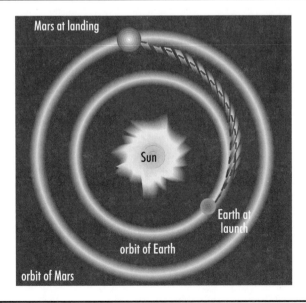

transit

In astronomy, the passage of a smaller object across the visible disk of a larger one. Transits of the inferior planets occur when they pass directly between Earth and the sun and are seen as tiny dark spots against the sun's disk. Other forms of transit include the passage of a satellite or its shadow across the disk of Jupiter and the passage of planetary-surface features across the central meridian of that planet as seen from Earth. The passage of an object in the sky across the observer's meridian is also known as a transit.

Triangulum

Small constellation of the Northern Hemisphere, represented as a triangle.

Triangulum Australe

Small bright constellation of the Southern Hemisphere, popularly known as Southern Triangle; it is more conspicuous than its northern counterpart, Triangulum.

Triton

The largest of Neptune's moons. It has a diameter of 2,700 km/1,680 mi, and orbits Neptune every 5.88 Earth days in a retrograde (east-to-west) direction. It takes the same time to rotate about its own axis as it does to make one revolution of Neptune.

It is slightly larger than the planet Pluto, which it is thought to resemble in composition and appearance. Probably, Triton was formerly a separate body like Pluto but was captured by Neptune. Triton was discovered in 1846 by British astronomer William Lassell only weeks after the discovery of Neptune. Triton's surface, as revealed by the U.S. *Voyager 2* space probe, has a temperature of 38 K (−235°C/−391°F), making it the coldest-known place in the solar system. It is covered with frozen nitrogen and methane, some of which evaporates to form a tenuous atmosphere with a pressure only 0.00001 that of Earth at sea level. Triton has a pink south polar cap, probably colored by the effects of solar radiation on methane ice. Dark streaks on Triton are thought to be formed by geysers of liquid nitrogen. The surface has few impact craters (the largest is the Mazomba, with a diameter of 27 km/17 mi), indicating that many of the earlier craters have been erased by the erupting and freezing of water (cryovulcanism).

Tucana

Constellation of the Southern Hemisphere, represented as a toucan. It contains the second-most-prominent globular cluster in the sky, 47 Tucanae, and the Small Magellanic Cloud.

Tunguska Event

Explosion at Tunguska, central Siberia, Russia, in June 1908 that devastated 6,500 km²/2,500 mi² of forest. It is thought to have been caused by either a cometary nucleus or a fragment of Encke's comet, about 200 m/660 ft across, or possibly an asteroid. The magnitude of the explosion was equivalent to an atomic bomb (10–

20 megatons) and produced a colossal shock wave; a bright falling object was seen 600 km/375 mi away and was heard up to 1,000 km/625 mi away.

An expedition to the site was made in 1927. The central area of devastation was occupied by trees that were erect but stripped of their branches. Farther out, to a radius of 20 km/12 mi, trees were flattened and laid out radially.

Tyuratam
Site of the Baikonur Cosmodrome, where the first satellite *Sputnik 1* was launched, in Kazakhstan. It is still operational today.

ultraviolet astronomy
Study of cosmic ultraviolet emissions using artificial satellites. The United States launched a series of satellites for this purpose, receiving the first useful data in 1968. Only a tiny percentage of solar ultraviolet radiation penetrates the atmosphere, this being the less dangerous longer-wavelength ultraviolet. The dangerous shorter-wavelength radiation is absorbed by the ozone layer high in Earth's upper atmosphere.

The U.S. *Orbiting Astronomical Observatory* (OAO) satellites provided scientists with a great deal of information regarding cosmic ultraviolet emissions. *OAO-1*, launched in 1966, failed after only three days, although *OAO-2*, put into orbit in 1968, operated for four years instead of the intended one year and carried out the first ultraviolet observations of a supernova and of Uranus. *OAO-3 (Copernicus)*, launched in 1972, continued transmissions into the 1980s and discovered many new ultraviolet sources. The *International Ultraviolet Explorer* (IUE), which was launched in January 1978 and ceased operation in September 1996, observed all of the main objects in the solar system, including Halley's comet, stars, galaxies, and the interstellar medium.

Ulysses
Space probe to study the sun's poles, launched in 1990 from the U.S. space shuttle *Discovery*. It is a joint project by U.S. National Aeronautics and Space Administration (NASA) and the European Space Agency. In February 1992, the gravity of Jupiter swung *Ulysses* onto a path that looped it first under the sun's south pole in 1994 and then over its north pole in 1995 to study the sun and solar wind at latitudes not observable from Earth.

United Kingdom Infrared Telescope (UKIRT)
A 3.8-m/150-in reflecting telescope for observing at infrared wavelengths, opened in 1979, on Mauna Kea, Hawaii, and operated by the Royal Observatory, Edinburgh, Scotland.

universal time (UT)
Another name for Greenwich Mean Time (GMT). It is based on the rotation of Earth, which is not quite constant. Since 1972, UT has been replaced by coordinated universal time (UTC), which is based on uniform atomic time.

universe

All of space and its contents, the study of which is called cosmology. The universe is thought to be between ten billion and twenty billion years old and is mostly empty space, dotted with galaxies for as far as telescopes can see. The most distant detected galaxies and quasars lie 10 billion light-years or more from Earth and are moving farther apart as the universe expands. Several theories attempt to explain how the universe came into being and evolved, such as the Big Bang theory of an expanding universe originating in a single explosive event and the contradictory steady-state theory.

Apart from those galaxies within the Local Group, all of the galaxies we see display red shifts in their spectra, indicating that they are moving away from us. The farther we look into space, the greater are the observed red shifts, which implies that the more distant galaxies are receding at ever greater speeds.

This observation led to the theory of an expanding universe, first proposed in 1929 by U.S. astronomer Edwin Hubble, and to Hubble's Law, which states that the speed with which one galaxy moves away from another is proportional to its distance from it. Current data suggest that the galaxies are moving apart at a rate of 50–100 kps/30–60 mps for every million parsecs of distance.

Uranus

The seventh planet from the sun, discovered in 1871 by German-born British astronomer William Herschel. It is twice as far out as the sixth planet, Saturn. Uranus has a mass 14.5 times that of Earth. The spin axis of Uranus is tilted at 98°, so that one pole points toward the sun, giving extreme seasons.

mean distance from the sun:
2.9 billion km/1.8 billion mi

equatorial diameter:
50,800 km/31,600 mi

rotation period:
.72 Earth days

year:
84 Earth years

atmosphere:
Deep atmosphere composed mainly of hydrogen and helium

surface:
Composed primarily of rock and various ices with only about 15 percent hydrogen and helium but may also contain heavier elements, which might account for Uranus's mean density being higher than Saturn's.

satellites:
Seventeen moons, the last two of which were discovered in 1997; Titania, the largest moon, has a diameter of 1,580 km/980 mi.

rings:
Eleven rings, composed of rock and dust, around the planet's equator were detected by the U.S. space probe *Voyager 2*. The rings are charcoal black and may be debris of former "moonlets" that have broken up. The ring farthest from the planet center (51,000 km/31,800 mi), Epsilon, is 100 km/62 mi at its widest point. In 1995, U.S. astronomers determined that the ring particles contained long-chain hydrocarbons. Looking at the brightest region of Epsilon, they were also able to calculate the precession of Uranus as 264 days, the fastest-known precession in the solar system.

Uranus has a peculiar magnetic field, in that it is tilted at 60° to the axis of spin and is displaced about a third of the way from the planet's center to its surface. Uranus spins from east to west, the opposite of the other planets, with the exception of Venus and possibly Pluto. The rotation rate of the atmosphere varies with latitude, from about sixteen hours in mid-southern latitudes to longer than seventeen hours at the equator.

Ursa Major
Latin for "Great Bear"; the third-largest constellation in the sky, in the north polar region. Its seven brightest stars make up the familiar shape or asterism of the Big Dipper or Plow. The second star of the handle of the dipper, called Mizar, has a companion star, Alcor. Two stars forming the far side of the dipper bowl act as pointers to the north Pole Star, Polaris. Dubhe, one of them, is the constellation's brightest star.
Many of the stars in this constellation are moving through space at the same speed and in the same direction. Sirius and several other stars in different parts of the sky share this common motion and so belong to the "Ursa Major moving group."

Ursa Minor
Latin for "Little Bear"; a small constellation of the Northern Hemisphere, popularly known as the Little Dipper. It is shaped like a dipper, with the bright north Pole Star, Polaris, at the end of the handle. Two other bright stars in this group, Beta and Gamma Ursae Minoris, are called "the Guards" or "the Guardians of the Pole." The constellation also contains the orange subgiant Kochab, about 95 light-years from Earth.

U.S. Naval Observatory
U.S. government observatory in Washington, D.C., which provides the nation's time service and publishes almanacs for navigators, surveyors, and astronomers. It contains a 66-cm/26-in refracting telescope, opened in 1873. It also operates a

1.55-m/61-in reflector for measuring positions of celestial objects, opened in 1964, at Flagstaff, Arizona.

Van Allen radiation belts

Two zones of charged particles around Earth's magnetosphere, discovered in 1958 by U.S. physicist James Van Allen. The atomic particles come from Earth's upper atmosphere and the solar wind and are trapped by Earth's magnetic field. The inner belt lies 1,000–5,000 km/620–3,100 mi above the equator and contains protons and electrons. The outer belt lies 15,000–25,000 km/9,300–15,500 mi above the equator but is lower around the magnetic poles. It contains mostly electrons from the solar wind. The Van Allen belts are hazardous to astronauts and interfere with electronic equipment on satellites.

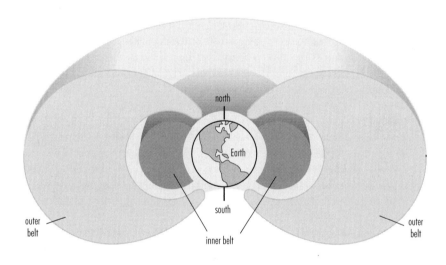

Vanguard

Early series of U.S. Earth-orbiting satellites and their associated rocket launcher. *Vanguard 1* was the second U.S. satellite, launched March 17, 1958, by the three-stage *Vanguard* rocket. Tracking of its orbit revealed that Earth is slightly pear shaped. The series ended in September 1959 with *Vanguard 3*.

variable star

A star whose brightness changes, either regularly or irregularly, over a period ranging from a few hours to months or years. The Cepheid variables regularly expand and contract in size every few days or weeks.

Stars that change in size and brightness at less precise intervals include long-period variables, such as the red giant Mira in the constellation Cetus (period about 331 days), and irregular variables, such as some red supergiants. Eruptive variables emit sudden outbursts of light. Some suffer flares on their surfaces, while others, such as a nova, result from transfer of gas between a close pair of stars. A supernova is the explosive death of a star. In an eclipsing binary, the variation is due not to any change in the star itself but to the periodic eclipse of a star by a close companion. The different types of variability are closely related to different stages of stellar evolution.

eclipsing binaries
Of nineteen thousand known variable stars, 80 percent are intrinsic variables and 20 percent are eclipsing binaries, pairs of stars in orbit around each other, whose combined light drops when one star eclipses the other. Much can be learned of the structure of the eclipsing system from detailed measurements of the light curve. In cases in which the radial velocities of both components can also be observed, it is possible to deduce the mass, diameter, and temperature of each star. The best-known eclipsing binary is Algol. Some eclipsing binaries consist of pairs of stars orbiting very close to each other or almost in contact, and such stars show severe deformation due to their mutual gravitational attraction. Orbital periods of eclipsing binaries range from about four hours to twenty-seven years.

intrinsic variable stars
Intrinsic variables fall into many classes, named for typical members. Regular or irregular pulsations account for the light variations of the majority of known variable stars, with periods ranging from minutes to years. In general, the light variation is greater in the stars of longer period. A numerous class of variable stars of short to moderate period consists of Cepheid variables, named for Delta Cephei. This star pulsates in a very regular period of 5.37 days, with a light variation of 0.8 magnitude, just more than twice as bright at maximum as at minimum. The periods of Cepheids range from three to fifty days, and there is a relationship between their period and their brightness: the longer the period, the brighter the Cepheid. Once the relationship had been established in 1912, the observed period of a very distant Cepheid could be used to determine its intrinsic luminosity and, hence, taking into account its apparent luminosity, to find its distance.

There are many variable red-giant stars, with periods ranging from thirty to a thousand days. Well-known examples are Betelgeuse and Mira. An interesting type of eruptive variable star is named for UV Ceti; such stars undergo small, sudden, irregular outbursts every few hours or days, with simultaneous emission of radio waves.

Vega, or Alpha Lyrae
Brightest star in the constellation Lyra and the fifth-brightest star in the night sky. It is a blue-white star, 25 light-years from Earth, with a luminosity fifty times that of the sun. In 1983, the *Infrared Astronomy Satellite* (IRAS) discovered a ring of

dust around Vega, possibly a disk from which a planetary system is forming. As a result of precession (the slow wobble of the Earth on its axis), Vega will become the north Pole Star about A.D. 14000.

Vela

Bright constellation of the Southern Hemisphere near Carina, represented as the sails of a ship. It contains large wisps of gas—called the Gum Nebula after its discoverer, Australian astronomer Colin Gum—believed to be the remains of one or more supernovae. Vela also contains the second optical pulsar (a pulsar that flashes at a visible wavelength) to be discovered. Vela was originally regarded as part of Argo. Its four brightest stars are second magnitude; one of them, Suhail, is about 490 light-years from Earth.

Venus

Second planet from the sun. It can approach Earth to within 38 million km/24 million mi, closer than any other planet. Its mass is 0.82 that of Earth. Venus rotates on its axis more slowly than any other planet, from east to west, the opposite direction to the other planets (except Uranus and possibly Pluto).

mean distance from the sun:
108.2 million km/67.2 million mi

equatorial diameter:
12,100 km/7,500 mi

rotation period:
243 Earth days

year:
225 Earth days

atmosphere:
Venus is shrouded by clouds of sulfuric acid droplets that sweep across the planet from east to west every four days. The atmosphere is almost entirely carbon dioxide, which traps the sun's heat by the greenhouse effect and raises the planet's surface temperature to 480°C/900°F, with an atmospheric pressure ninety times that at the surface of Earth.

surface:
Consists mainly of silicate rock and may have an interior structure similar to that of Earth: an iron–nickel core, a mantle composed of more mafic rocks (rocks made of one or more ferromagnesian, dark-colored minerals), and a thin siliceous outer crust. The surface is dotted with deep impact craters. Some of Venus's volcanoes may still be active. The largest highland area is Aphrodite Terra near the equator, half the size of Africa. The highest mountains are on

the northern highland region of Ishtar Terra, where the massif of Maxwell Montes rises to 10,600 m/35,000 ft above the average surface level. The highland areas on Venus were formed by volcanoes.

satellites:
No moons

The first artificial object to hit another planet was the Soviet probe *Venera 3,* which crashed on Venus on March 1, 1966. Later *Venera* probes parachuted down through the atmosphere and landed successfully on its surface, analyzing surface material and sending back information and pictures. In December 1978 a U.S. *Pioneer-Venus* probe went into orbit around the planet and mapped most of its surface by radar, which penetrates clouds. In 1992, the U.S. space probe *Magellan* mapped 99 percent of the planet's surface to a resolution of 100 m/330 ft.

Venus has an ion-packed tail 45 million km/28 million mi in length that stretches away from the sun and is caused by the bombardment of the ions in Venus's upper atmosphere by the solar wind. It was first discovered in the late 1970s but it was not until 1997 that the *Solar and Heliospheric Observatory* (SOHO) revealed its immense length.

vernal equinox
Spring equinox. The point during spring when the sun's path crosses the equator so that day and night are approximately of equal length.

Very Large Array (VLA)
Largest and most complex single-site radio telescope in the world. It is located on the Plains of San Augustine, west of Socorro, New Mexico. It consists of twenty-seven dish antennae, each 25 m/82 ft in diameter, arranged along three equally spaced arms forming a Y-shaped array. Two of the arms are 21 km/13 mi long, and the third, to the north, is 19 km/11.8 mi long. The dishes are mounted on railroad tracks, enabling the configuration and size of the array to be altered as required.

Pairs of dishes can also be used as separate interferometers, each dish having its own individual receivers that are remotely controlled, enabling many different frequencies to be studied. There are four standard configurations of antennae, ranging from A (the most extended) through B and C to D. In the A configuration, the antennae are spread out along the full extent of the arms, and the VLA can map small, intense radio sources with high resolution. The smallest configuration, D, uses arms that are just 0.6 km/0.4 mi long for mapping larger sources. Here, the resolution is lower, although there is greater sensitivity to fainter, extended fields of radio emission.

Very Long Baseline Array (VLBA)
A group of ten 25-m/82-ft radio telescopes spread across North America and Hawaii that operate as a single instrument, using the technique of very long

baseline interferometry (VBLI). The longest baseline (distance between pairs of telescopes) is about 8,000 km/4,970 mi.

very long baseline interferometry (VLBI)

In radio astronomy, a method of obtaining high-resolution images of astronomical objects by combining simultaneous observations made by two or more radio telescopes thousands of kilometers apart. The maximum resolution that can be achieved is directly proportional to the longest baseline in the array (the distance between any pair of telescopes) and inversely proportional to the radio wavelength being used.

Viking probes

Two U.S. space probes to Mars, each one consisting of an orbiter and a lander. Launched on August 20 and September 9, 1975, respectively, they transmitted color pictures and analyzed the soil. *Viking 1* carried life-detection labs and landed in the Chryse lowland area on July 20, 1976, for detailed research and photos. Designed to work for ninety days, it operated for 6.5 years, going silent in November 1982. *Viking 2* was similar in setup to *Viking 1;* it landed in Utopia on September 3, 1976, and functioned for 3.5 years.

Virgo

Zodiacal constellation of the Northern Hemisphere, the second-largest in the sky. It is represented as a maiden holding an ear of wheat, marked by first-magnitude Spica, Virgo's brightest star. The sun passes through Virgo from late September to the end of October. In astrology, the dates for Virgo are about August 23–September 22. Virgo contains the nearest large cluster of galaxies to us, 50 million light-years away, consisting of about three thousand galaxies centered on the giant elliptical galaxy M87. Also in Virgo is the nearest quasar, 3C 273, an estimated 3 billion light-years away.

VLA

See Very Large Array (VLA).

VLBA

See Very Long Baseline Array (VLBA).

VLBI

See very long baseline interferometry (VLBI).

Volans

Small constellation of the Southern Hemisphere, represented as a flying fish. It was formerly known as Piscis Volans.

Voskhod

Russian for "ascent"; a Soviet spacecraft used in the mid-1960s; it was modified from the single-seat *Vostok* and was the first spacecraft capable of carrying two or three cosmonauts. During *Voskhod 2's* flight in 1965, Aleksi Leonov made the first space walk.

Vostok

Russian for "east"; the first Soviet spacecraft, used 1961–1963. *Vostok* was a metal sphere 2.3 m/7.5 ft in diameter, capable of carrying one cosmonaut. It made flights lasting up to five days. In April 1961 *Vostok 1* carried the first person into space, Yuri Gagarin.

Voyager probes

Two U.S. space probes. *Voyager 1,* launched September 5, 1977, passed Jupiter in March 1979 and reached Saturn in November 1980. *Voyager 2* was launched earlier (August 20, 1977) on a slower trajectory that took it past Jupiter in July 1979, Saturn in August 1981, Uranus in January 1986, and Neptune in August 1989. Like the *Pioneer* probes, the *Voyagers* are on their way out of the solar system; at the start of 1995, *Voyager 1* was 8.8 billion km/5.5 billion mi from Earth, and *Voyager 2* was 6.8 billion km/4.3 billion mi from Earth. Their tasks now include helping scientists locate the position of the heliopause, the boundary at which the influence of the sun gives way to the forces exerted by other stars. Both *Voyagers* carry specially coded long-playing records called *Sounds of Earth* for the enlightenment of any other civilizations that might find them.

Vulpecula

Small constellation in the Northern Hemisphere just south of Cygnus, represented as a fox. It contains a major planetary nebula, the Dumbbell, and the first pulsar (pulsating radio source) to be discovered.

white dwarf

Small, hot star, the last stage in the life of a star such as the sun. White dwarfs make up 10 percent of the stars in the galaxy; most have a mass 0.6 that of the sun, but a diameter only 0.01 that of the sun, similar in size to Earth. Most have surface temperatures of 8,000°C/14,400°F or more, hotter than the sun. Yet, because they are so small, their overall luminosities may be less than 1 percent that of the sun. The Milky Way contains an estimated fifty billion white dwarfs.

White dwarfs consist of degenerate matter in which gravity has packed the protons and electrons together as tightly as is physically possible, so that a spoonful of it weighs several metric tons. White dwarfs are thought to be the shrunken remains of stars that have exhausted their internal energy supplies. They slowly cool and fade over billions of years.

WIMP (weak interacting massive particle)

Hypothetical subatomic particle found in the galaxy's dark matter. These particles could constitute the 80 percent of dark matter unaccounted for by MACHOs (massive astrophysical compact halo objects).

X-ray astronomy

Detection of X-rays from intensely hot gas in the universe. Such X-rays are prevented from reaching Earth's surface by the atmosphere, so detectors must be placed in rockets and satellites. The first celestial X-ray source, Scorpius X-1, was discovered in by a rocket flight. Since 1970, special satellites have been put into orbit to study X-rays from the sun, stars, and galaxies. Many X-ray sources are believed to be gas falling onto neutron stars and black holes.

X-ray telescope

A telescope designed to receive electromagnetic waves in the X-ray part of the spectrum. X-rays cannot be focused by lenses or mirrors in the same way as visible light, and numerous alternative techniques are used to form images. Because X-rays cannot penetrate Earth's atmosphere, X-ray telescopes are mounted on satellites, rockets, or high-flying balloons.

year

Unit of time measurement, based on the orbital period of Earth around the sun. The tropical year (also called equinoctial and solar year), from one spring equinox to the next, lasts 365.2422 days. It governs the occurrence of the seasons and is the period on which the calendar year is based. The sidereal year is the time taken for Earth to complete one orbit relative to the fixed stars and lasts 365.26 days (about twenty minutes longer than a tropical year). The difference is due to the effect of precession, which slowly moves the position of the equinoxes. The anomalistic year is the time taken by any planet in making one complete revolution from perihelion to perihelion; for Earth, this period is about five minutes longer than the sidereal year due to the gravitational pull of the other planets. The calendar year consists of 365 days, with an extra day added at the end of February each leap year. Leap years occur in every year that is divisible by four, except that a century year is not a leap year unless it is divisible by 400. Hence, 1900 was not a leap year but 2000 will be.

Yerkes Observatory

Astronomical center in Williams Bay, Wisconsin, founded by U.S. astronomer George Ellery Hale in 1897. It houses the world's largest refracting optical telescope, with a lens of diameter 102 cm/40 in.

Zelenchukskaya

Site of the world's largest single-mirror optical telescope, with a mirror of 6-m/ 19.7-ft diameter, in the Caucasus Mountains of Russia. At the same site is the RATAN 600 radio telescope, consisting of radio reflectors in a circle of 600-m/ 2,000-ft diameter. Both instruments are operated by the Academy of Sciences in St. Petersburg.

zenith

Uppermost point of the celestial horizon, immediately above the observer; the nadir is below, diametrically opposite.

zodiac

Zone of the heavens containing the paths of the sun, the moon, and the planets. When this was devised by the ancient Greeks, only five planets were known, making the zodiac about 16° wide. In astrology, the zodiac is divided into twelve signs, each 30° wide: Aries, Taurus, Gemini, Cancer, Leo, Virgo, Libra, Scorpio, Sagittarius, Capricorn, Aquarius, and Pisces. These do not cover the same areas of sky as the astronomical constellations. The twelve astronomical constellations are uneven in size and do not, among them, cover the whole zodiac, or even the line of the ecliptic, much of which lies in the constellation of Ophiuchus.

The idea of such a zone is ancient, but the present name is of Greek origin. At first, various asterisms were chosen along the zodiac to serve as calendar reference points, but, as these were unequally spaced, the zone was eventually divided into twelve equal signs, each 30° wide. The sequence of the signs is eastward, following the motions of the sun and the moon through the constellations, and is regarded as beginning at the point that marks the position of the sun at the time of the March equinox. This point is sometimes called "the First Point of Aries."

The equinoxes occur as the sun enters the signs of Aries and Libra, at which times the sun passes directly above the equator. The solstices occur as the sun enters the signs of Cancer and Capricorn and is passing directly over the Tropic of Cancer and the Tropic of Capricorn, circles of latitude 23°27' north and south of the equator.

Because of precession (the slow wobble of the Earth on its axis), the equinoctial point—and, with it, the zodiacal signs—moves westward through the constellations at a rate of one sign in 2,150 years. The First Point of Aries, which is now in Pisces, would, at the time of Greek astronomer Hipparchusin in the second century B.C., have been in Aries. At the time the present constellation groupings were first made, two thousand years earlier, the First Point of Aries would have been in Taurus, and the initial asterism of the zodiac would have been the Pleiades.

zodiacal light

Cone-shaped light sometimes seen extending from the sun along the ecliptic (i.e., the path that the sun appears to follow each year as it is orbited by Earth), visible after sunset or before sunrise. It is due to thinly spread dust particles in the central plane of the solar system. It is very faint and requires a dark, clear sky to be seen. The dust is thought to be produced mainly by colliding asteroids and active comets, though a small amount probably originates from outside the solar system. Astronomers estimate that 10,000 metric tons/11,023 tons of this dust hits Earth each year.

Astronomical Constants

Constant	Value
Astronomical unit (au)	149.6 million km/92.96 million mi
Speed of light in a vacuum (c)	299,792.458 km/sec
Solar parallax	8.794148 arc seconds
Mass of the sun	1.9891×10^{30} kg
Mass of the earth	5.9742×10^{24} kg
Mass of the moon	7.3483×10^{22} kg
Light-year (ly)	9.4605×10^{12} km (or 0.30660 pc)
Parsec (pc)	30.857×10^{12} km (or 3.2616 ly)
Obliquity of the elliptic (2000)	23° 26' 21.448"
General precession (2000)	50.290966 arc seconds/year
Constant of nutation (2000)	9.2025 arc seconds
Constant of aberration (2000)	20.49552 arc seconds

Constellations

Constellation	Abbreviation	Popular name(s)
Andromeda	And	Andromeda
Antlia	Ant	Air Pump
Apus	Aps	Bird of Paradise
Aquarius	Aqr	Water-Bearer
Aquila	Aqi	Eagle
Ara	Ara	Altar
Aries	Ari	Ram
Auriga	Aur	Charioteer
Boötes	Boo	Herdsman
Caelum	Cae	Chisel
Camelopardalis	Cam	Giraffe
Cancer	Cnc	Crab
Canes Venatici	CVn	Hunting Dogs
Canis Major	CMa	Great Dog
Canis Minor	CMi	Little Dog
Capricornus	Cap	Sea-Goat
Carina	Car	Keel
Cassiopeia	Cas	Cassiopeia
Centaurus	Cen	Centaur
Cepheus	Cep	Cephus
Cetus	Cet	Sea Monster, Whale
Chamaeleon	Cha	Chameleon

Constellation	Abbreviation	Popular name(s)
Circinus	Cir	Compasses
Columba	Col	Dove
Coma Berenices	Com	Berenice's Hair
Corona Australis	CrA	Southern Crown
Corona Borealis	CrB	Northern Crown
Corvus	Crv	Crow
Crater	Crt	Cup
Crux	Cru	Southern Cross
Cygnus	Cyn	Swan
Delphinus	Del	Dolphin
Dorado	Dor	Goldfish, Swordfish
Draco	Dra	Dragon
Equuleus	Equ	Foal
Eridanus	Eri	River
Fornax	For	Furnace
Gemini	Gem	Twins
Grus	Gru	Crane
Hercules	Her	Hercules
Horologium	Hor	Clock
Hydra	Hya	Sea Serpent
Hydrus	Hyi	Watersnake
Indus	Ind	Native American
Lacerta	Lac	Lizard
Leo	Leo	Lion
Leo Minor	LMi	Little Lion
Lepus	Lep	Hare
Libra	Lib	Scales
Lupus	Lup	Wolf
Lynx	Lyn	Lynx
Lyra	Lyr	Harp

Constellation	Abbreviation	Popular name(s)
Mensa	Men	Table, Mountain
Microscopium	Mic	Microscope
Monoceros	Mon	Unicorn
Musca	Mus	Fly
Norma	Nor	Level, Square
Octans	Oct	Octant
Ophiuchus	Oph	Serpent-Bearer
Orion	Ori	Orion
Pavo	Pav	Peacock
Pegasus	Peg	Flying Horse
Perseus	Per	Perseus
Phoenix	Phe	Phoenix
Pictor	Pic	Painter, Easel
Pisces	Psc	Fishes
Piscis Austrinus	PsA	Southern Fish
Puppis	Pup	Poop, Stern
Pyxis	Pyx	Compass
Reticulum	Ret	Net
Sagitta	Sge	Arrow
Sagittarius	Sgr	Archer
Scorpius	Sco	Scorpion
Sculptor	Scl	Sculptor's Tools
Scutum	Sct	Shield
Serpens	Ser	Serpent
Sextans	Sex	Sextant
Taurus	Tau	Bull
Telescopium	Tel	Telescope
Triangulum	Tri	Triangle
Triangulum Australe	TrA	Southern Triangle
Tucana	Tuc	Toucan

Constellation	Abbreviation	Popular name(s)
Ursa Major	UMa	Big Dipper
Ursa Minor	UMi	Little Dipper
Vela	Vel	Sail
Virgo	Vir	Virgin
Volans	Vol	Flying Fish
Vulpecula	Vul	Fox

The Planets
(by Distance from the Sun)

Planet	Constituents	Atmosphere	Distance from sun km/mi (000,000)	Orbit (Earth years)	Diameter km/mi (000)	Density (water = 1 unit)
Mercury	rock, ferrous	minute traces of argon, helium	58/36	0.241	4.88/3.03	5.4
Venus	rock, ferrous	carbon dioxide	108.2/67.20	0.616	12.10/7.51	5.2
Earth	rock, ferrous	nitrogen, oxygen	149.5/92.90	1.00	12.76/7.92	5.5
Mars	rock	carbon dioxide	227.90/141.60	1.88	6.78/4.21	3.9
Jupiter	liquid hydrogen, helium	—	778/483	11.86	142.80/88.70	1.3
Saturn	hydrogen, helium	—	1,427/886	29.46	120.00/75.00	0.7
Uranus	ice, hydrogen, helium	hydrogen, helium	2,900/1,800	84.00	50.80/31.60	1.3
Neptune	ice, hydrogen, helium	hydrogen helium	4,400/2,794	164.80	48.60/30.20	1.8
Pluto	ice, rock	methane	5,800/3,600	248.50	2.30/1.43	~2

Largest Natural Planetary Satellites

Planet	Satellite	Diameter km/mi	Mean distance from center of primary planet km/mi	Orbital period (Earth days)	Reciprocal mass (planet = 1)
Jupiter	Ganymede	5,262/3,300	1,070,000/664,898	7.16	12,800
Saturn	Titan	5,150/3,200	1,221,800/759,226	15.95	4,200
Jupiter	Callisto	4,800/3,000	1,883,000/1,170,096	16.69	17,700
Jupiter	Io	3,630/2,240	421,600/261,982	1.77	21,400
Earth	Moon	3,476/2,160	384,400/238,866	27.32	81
Jupiter	Europa	3,138/1,900	670,900/416,897	3.55	39,700
Neptune	Triton	2,700/1,690	354,300/220,162	5.88	770

Star: Nearest Stars

Star	Distance (light-years)
Proxima Centauri	4.2
Alpha Centauri A	4.3
Alpha Centauri B	4.3
Barnard's Star	6.0
Wolf 359	7.7
Lalande 21185	8.2
UV Ceti A	8.4
UV Ceti B	8.4
Sirius A	8.6
Sirius B	8.6
Ross 154	9.4
Ross 249	10.4

The 20 Brightest Stars

Scientific name	Common name	Distance (light-years)
Alpha Canis Majoris	Sirius	8.6
Alpha Carinae	Canopus	1,170
Alpha Centauri	Rigil Kent	4.3
Alpha Boötis	Arcturus	36
Alpha Lyrae	Vega	25
Alpha Aurigae	Capella	42
Beta Orionis	Rigel	910
Alpha Canis Minoris	Procyon	11.4
Alpha Eridani	Achernar	144
Alpha Orionis	Betelgeuse	310
Beta Centauri	Hadar	320
Alpha Aquilae	Altair	16
Alpha Crucis	Acrux	360
Alpha Tauri	Aldebaran	65
Alpha Virginis	Spica	274
Alpha Scorpii	Antares	420
Beta Geminorum	Pollux	36
Alpha Piscis Austrini	Fomalhaut	25
Alpha Cygni	Deneb	1,400
Beta Crucis	Mimosa	420

Major Ground-Based Telescopes and Observatories

Observatory/ Telescope	Location	Description	Year opened	Run by
Algonquin Radio Observatory	Ontario, Canada	radio telescope, 46 m/150 ft diameter	1966	National Research Council (NRC) of Canada
Apache Point Observatory	Sacramento Mountains, New Mexico	3.5-m/138-in diameter optical/ infrared telescope	1990	New Mexico State University, University of Washington, University of Chicago, Princeton University, Washington State University
Arecibo Observatory	Puerto Rico	home of the largest single-dish radio telescope in the world; a 305-m-/1,000-ft-diameter spherical reflector with a surface made up of nearly 40,000 perforated aluminum panels. Each panel can be adjusted to maintain a precise spherical shape that varies less than 3 mm/0.12 in over the whole 80,937-sq m/ 20-acre surface	inaugurated in 1963; upgraded in 1974, and twice in the mid-1990s	National Astronomy and Ionosphere Center, which is operated by Cornell University and the National Science Foundation
Australia Telescope National Facility	New South Wales, Australia	giant radio telescope consisting of: six-m/72-ft antennae at Culgoora, near Narrabri, a similar antenna at Mopra, near Siding Spring	1989	Commonwealth Scientific and Industrial Research Organization (CSIRO)

Observatory/ Telescope	Location	Description	Year opened	Run by
		Mountain, and the 64-m/ 210-ft Parkes radio telescope; together they simulate a dish 300 km/186 mi across		
Cerro Tololo Inter-American Observatory	Cerro Tololo Mountain in the Chilean Andes	main instrument is a 4-m/158-in reflector, a twin of that at Kitt Peak, Arizona	1974	Association of Universities for Research into Astronomy (AURA)
David Dunlap Observatory	Richmond Hill, Ontario, Toronto	main instrument is 1.88-m/74-in reflector, the largest optical telescope in Canada	1935	University of Canada
Dominion Astrophysical Observatory	near Victoria, British Columbia, Canada	1.85-m/73-in reflector, and 1.2-m/48-in reflector	1918	National Reserch Council (NRC) of Canada
Dominion Radio Astrophysical Observatory	Penticton, British Columbia, Canada	26-m/84-ft radio dish and an aperture synthesis radio telescope	1959	National Reserch Council (NRC) through its Herzborg Institute of Astrophysics
Effelsberg Radio Telescope	near Bonn, Germany	the world's largest fully steerable radio telescope; 100-m/328-ft radio dish	1971	Max Planck Institute for Radio Astronomy
European Southern Observatory	La Silla, Chile	telescopes include: 3.6 m/ 142 in reflector	1976	operated jointly by Belgium, Denmark, France, Germany, Italy, the Netherlands, Sweden, and Switzerland with headquarters near Munich, Germany
		3.5-m/138-in New Technology Telescope	1990	
	Cerro Paranal, Chile	Very Large Telescope (VLT), consisting of four 8-m/26-ft reflectors and several 1.8-m/70.8-in telescopes mounted independently but capable of working in combination	2001	

Observatory/ Telescope	Location	Description	Year opened	Run by
Gemini 8-Meter Telescopes Project	Mauna Kea, Hawaii; Cerro Pachon, Chile	two 8-m/26-ft aperture optical/infared telescopes	1999-2000	international partnership of United States, Britain, Canada, Chile, Australia, Argentina, and Brazil
Green Bank Telescope	Green Bank, Pocahontas County, West Virginia	largest fully steerable radio telescope in the world with a 100-m x 110-m/300-ft x 340-ft surface	1999	National Radio Astronomy Observatory (NRAO)
Jodrell Bank	Cheshire, UK	home of two telescopes: Lovell Telescope, a 76-m/ 250-ft radio dish	1957, modified in 1970	Nuffield Radio Astronomy Laboratories of the University of Manchester
		an elliptical radio dish 38 m × 25 m/125 ft × 82 ft capable of working at shorter wavelengths	1964	
Keck I	Mauna Kea, Hawaii	world's largest optical telescope, with a primary mirror 10-m/33-ft diameter, with 36 hexagonal sections, each controlled by a computer to generate single images of the objects observed	1992	California Institute of Technology and the University of California/Lick Observatory
Keck II	Mauna Kea, Hawaii	identical to Keck I	1996	California Institute of Technology and the University of California/Lick Observatory
Kitt Peak National Observatory	Quinlan Mountains near Tucson, Arizona	numerous telescopes including the 4-m/158-in Mayall reflector	1973	Association of Universities for Research (AURA) in agreement with the National Science Foundation of the United States

Observatory/ Telescope	Location	Description	Year opened	Run by
		McMath-Pierce Solar Telescope, the world's largest of its type, it comprises three telescopes: one 1.61-m/ 63-in and two 0.91-m/36-in	1962	National Solar Observatory (NSO)
		3.5-m/138-in reflecting telescope	1994	WIYN consortium that comprises the University of Wisconsin, Indiana University, Yale University, and National Optical Astronomy Observatories (NOAO)
La Palma Observatory (Observatorio del Roque de los Muchachos)	La Palma, Canary Islands	Isaac Newton Group of telescopes, including the 4.2-m/165-in William Herschel Telescope	1987	Particle Physics and Astronomy Research Council (PPARC) of Britain, Royal Observatory, Scotland
Las Campanas Observatory	Las Campanas, Chile	2.5-m/100-in Irénée du Pont telescope	1977	Carnegie Institution of Washington, D.C.
		Magellan I and II: two 6.5-m/21.3-ft honeycomb-back optical mirror telescopes	1999 Magellan I	Carnegie Institution of Washington, D.C., University of Arizona, Harvard University, Massachusetts Institute of Technology, and University of Michigan
Lick Observatory (the University of California/Lick Observatory)	Mount Hamilton, California	several instruments including: 3.04-m/120-in Shane reflector	1959	University of California
		91-cm/36-in refractor, the second-largest refractor in the world	1988	
Lowell Observatory	Flagstaff, Arizona	eight telescopes, including the 61-cm/24-in Alvan Clark refractor	1896	Lowell Observatory staff

Observatory/ Telescope	Location	Description	Year opened	Run by
	Anderson Mesa, Arizona	several telescopes, including the 1.83 m/72 in Perkins reflector		Ohio State and Ohio Wesleyan Universities
McDonald Observatory	Mount Locke, Davis Mountains, Texas	2.72-m/107-in reflector	1969	University of Texas
		Otto Struve telescope: 2.08-m/82-in reflector	1939	
		9.2-m/30.2-ft Hobby-Eberly Telescope (HET) for spectral analysis	1997	Penn State University, Stanford University, University of Texas at Austin, George August Universität, and the Ludwig-Maximilians Universität
Magellan I and II	Las Campanas, Chile	two 6.5-m/21.3-ft honeycomb-back optical mirror telescopes	1999 Magellan I	Carnegie Institution of Washington, D.C., University of Arizona, Harvard University, Massachusetts Institute of Technology, and University of Michigan
Mauna Kea	Mauna Kea, Hawaii	telescopes include: the 2.24-m/88-in University of Hawaii reflector	1970	University of Hawaii/ National Aeronautics and Space Administration (NASA)
		3.8-m/150-in United Kingdom Infrared Telescope (UKIRT) (also used for optical observations)	1979	Royal Observatory, Edinburgh, Scotland
		3-m/120-in NASA Infrared Telescope Facility (IRTF)	1979	National Aeronautics and Space Administration (NASA)

Observatory/ Telescope	Location	Description	Year opened	Run by
		3.6-m/142-in Canada-France-Hawaii Telescope (CFHT), designed for optical and infrared work	1979	National Research Council (NRC) of Canada, University of Hawaii, and Center National de la Recherche Scientifique, France
		15-m/50-ft diameter UK/Netherlands James Clerk Maxwell Telescope (JCMT), the world's largest telescope specifically designed to observe millimeter wave radiation from nebulae, stars, and galaxies	1987, enhanced 1996	Joint Astronomy Center in Hilo, Hawaii, for the National Research Council (NRC) of Canada, Particle Physics and Astronomy Research Council (PPARC) of Britain, and the Netherlands Organization for Scientific Research (NOSR)
		Keck I and Keck II, two 10-m/33-ft diameter optical telescopes	1992 and 1996	California Institute of Technology and the University of California/Lick Observatory
Mount Palomar	near Pasadena, Callifornia	5-m/200-in Hale reflector; 1.2-m/48-in Schmidt telescope; it was the world's premier observatory during the 1950s	1948	California Institute of Technology and the University of California/Lick Observatory
Mount Wilson Observatory	San Gabriel Mountains near Los Angeles California	several telescopes including: 2.5-m/100-in Hooker telescope, with which Edwin Hubble discovered the expansion of the universe	1917	Mount Wilson Institute
		1.5-m/60-in reflector telescope	1908	

Observatory/ Telescope	Location	Description	Year opened	Run by
		two solar telescopes in towers 18.3 m/60 ft and 45.7 m/150 ft tall	1912	University of California (UCLA) and University of Southern California
Mullard Radio Astronomy Observatory	Cambridge, England	Ryle Telescope, eight dishes 12.8 m/42 ft wide in a line 5 km/3 mi long	1972	University of Cambridge, England
Multiple Mirror Telescope	Mt Hopkins, Arizona	6.5-m/21.3-ft honeycomb-back optical telescope conversion (previously six 1.8-m/72-in mirrors which performed as a single 4.5-m/176-in mirror)	1996	University of Arizona and the Smithsonian Institution
New Technology Telescope	La Silla, Chile	optical telescope that forms part of the European Southern Observatory; it has a thin, lightweight mirror, 3.5 m/138 in across with active optics, which is kept in shape by computer-adjustable supports	1990	European Southern Observatory, consisting of eight European countries
Anglo-Australian Observatory	Siding Spring Mountain, New South Wales Australia	1.2-m/48-in UK Schmidt Telescope	1973	Anglo-Australian Telescope Board
		3.9-m/154-in Anglo-Australian Telescope	1975	
South African Astronomical Observatory	Sutherland, South Africa	main telescope is a 1.88-m/74-in reflector	founded 1973	Council for Scientific and Industrial Research of South Africa
Special Astrophysical Observatory	Zelenchukskaya, near the Caucasus Mountains of Russia	site of the world's largest single mirror optical telescope, with a mirror of 6-m/236-in diameter	1976	Russian Academy of Sciences (RAS), St. Petersburg
		Radio Astronomy Telescope of the Russian Academy of Sciences (RATAN) 600 radio telescope, consisting of radio reflectors in a circle 600 m/2,000 ft in diameter		

Observatory/ Telescope	Location	Description	Year opened	Run by
Subaru ("Pleiades")	Mauna Kea, Hawaii	8-m/26.4-ft optical-infrared telescope	2000	National Astronomical Observatory (NAO) of the University of Tokyo
United Kingdom Infrared Telescope (UKIRT)	Mauna Kea, Hawaii	3.8-m/150-in reflecting telescope for observing at infrared wavelengths	1979	Royal Observatory, Scotland
U.S. Naval Observatory	Washington, D.C.	several telescopes including: 66-cm/26-in refracting telescope	1873	U.S. Naval Observatory
	Flagstaff, Arizona	1.55-m/61-in reflector for measuring positions of celestial objects	1964	
Very Large Array (VLA)	Plains of San Augustine, New Mexico	largest and most complex single-site radio telescope in the world, comprising 27 dish antennae, each 25-m/82-ft in diameter, forming a Y-shaped array	1980	National Radio Astronomy Observatory (NRAO)
Very Large Telescope (VLT)	Cerro Paranal, Chile	four 8-m/26-ft reflectors in an optical array	2001	European Southern Observatory consisting of eight European countries
Very Long Baseline Array (VLBA)	St. Croix, Virgin Islands; Hancock, New Hampshire; North Liberty, Iowa; Fort Davis, Texas; Los Alamos, New Mexico; Pie Town, New Mexico; Kitt Peak, Arizona; Owens Valley, California; Brewster, Washington; Mauna Kea, Hawaii	system of 10 radio telescopes, each a 25-m/82-ft diameter dish antenna, controlled remotely from the Array Operations Center in Socorro, New Mexico, that work together as the world's largest dedicated, full-time astronomical instrument	1993	National Radio Astronomy Observatory (NRAO)

Observatory/ Telescope	Location	Description	Year opened	Run by
Yerkes Observatory	Williams Bay, Wisconsin	houses the world's largest refracting optical telescope, with a lens of diameter 102 cm/40 in	observatory founded 1897	University of Chicago, Department of Astronomy and Astrophysics

Index

[Note: page numbers in *italics* refer to illustrations.]